隈研吾的材料研究室

Kengo Kuma: a LAB for materials

[日]隈研吾 著

日本株式会社新建筑社 编

陆宇星 谭露 译

中信出版集团 | 北京

图书在版编目（CIP）数据

隈研吾的材料研究室/（日）隈研吾著；日本株式
会社新建筑社编；陆宇星，谭露译.-- 北京：中信出
版社，2020.1（2021.10重印）
书名原文：Kengo Kuma：a LAB for materials
ISBN 978-7-5217-0959-9

Ⅰ.①隈… Ⅱ.①隈… ②日… ③陆… ④谭… Ⅲ.
①建筑材料－研究 Ⅳ.① TU5

中国版本图书馆 CIP 数据核字（2019）第 177346 号

Kengo Kuma：a LAB for materials by Kengo Kuma
Copyright © 2018 Shinkenchiku-sha Co., Ltd.
《隈研吾的材料研究室》译自日本新建筑社出版的 JA109 号
日文版
发行人：吉田信之
编　辑：藤田惠弥子
编　务：藤原彻平
Simplified Chinese translation copyright © 2019 by CITIC Press Corporation
All rights reserved.

本书仅限中国大陆地区发行销售

隈研吾的材料研究室

著　　者：[日]隈研吾
编　　者：日本株式会社新建筑社
译　　者：陆宇星　谭露
出版发行：中信出版集团股份有限公司
　　　　　（北京市朝阳区惠新东街甲 4 号富盛大厦 2 座　邮编　100029）
承 印 者：北京利丰雅高长城印刷有限公司

开　　本：787mm×1092mm　1/16　　印　张：18.25　　字　数：250 千字
版　　次：2020 年 1 月第 1 版　　　　印　次：2021 年 10 月第 4 次印刷
京权图字：01-2019-6988
书　　号：ISBN 978-7-5217-0959-9
定　　价：168.00 元

目 录

Contents

回到材料
Return to Materials

隈研吾

20世纪的割裂

遇到新的材料，新的时代就开始了。原因很简单，建筑是用材料做出来的，材料决定建筑。尽管如此，20世纪却是一个低估了材料的价值的时代。导致人与材料完全割裂的原因有两个。

一个原因是这个时代被混凝土材料彻底覆盖，谁都理所当然地认为建筑就是用混凝土做的，因此没人愿意与其他材料打交道。

混凝土是非常适合20世纪的材料，这个时代的建筑需要被快速、大量地建造。人们完全忘记了过去是怎么做建筑的，争先恐后地奔向混凝土。

空间的时代

另一个原因是人们认为建筑即是空间，就是墙和柱子之间的空隙。这个席卷了20世纪的观点，是对此前世界观的一种批判。20世纪以前的建筑观认为，建筑就是物体本身，无外乎墙和柱子。而现代的建筑观认为，建筑最重要的是空隙的部分，即物体与物体之间的空间。这种观点很有说服力，它否定了分隔空间的厚重的墙体。在追求外部空间与内部空间、内部空间相互之间的自由连接的20世纪，建筑即空间这个定义是与之完美契合的。

牛顿以来的现代科学的世界观，对这种观念的形成也有很大的影响。按照牛顿的世界观，物体是在绝对、客观的空间内，在科学规律的支配下相互作用的。这个观点打开了一扇门，使我们进入了以空间为主角的世界。然而，如此定义世界，也让我们失去了很多东西。在建筑中，比起地板、墙和天花板，空气更重要，这样的想法一旦被接受，用手触摸墙壁、用脚底感知地板这些事情就被忽视了。建筑与人之间的关系变得淡薄，比以往任何时候都要弱。

由于建筑被定义为空间，人们开始仅以面积来评价建筑、计算建筑的价值：容积率是多少，一平方米的工程费是多少，租金是多少，等等。关于建筑的话题中，充斥着总建筑面积、建筑密度、容积率等术语，这都是因为物质被轻视，空气成了主角。这个贫瘠空虚的时代，持续了约一个世纪。

我想回归物质，想重建物质与人的联系。在过去的30年里，我一直在做这样的尝试。一旦上手，我就发现再没有比这更令人愉快的事情了。世界是由各种各样的物质构成的，未知的物质实在太多了。

物质与场所

有趣的是，物质与场所总有着密不可分的联系。一小片木头里也藏有一个无穷深刻、层次丰富的故事，比如树木生长的气候，谁砍掉了这棵树，他的文化背景怎样，他又是如何使用这些木头的。每一种物质都与它所在的场所联系在一起。以物质为媒介，我们可以了解它所在的场所，并掌握这个场所的具体信息。

我不相信不以物质为媒介的场所论。民族主义正是一种抽象又空洞的场所论的典型。从这个意义上说，我毫无疑问是一个唯物主义者。在我看来，20世纪是一个既遗忘了物质又忽视、低估了场所的世纪，人的脚离开了场所，离开了大地，只是漫无目的地游荡彷徨。

挪亚的洪水

在过去的100年里，人们完全忽视了思考和探索怎样用物质来建造建筑、建造城市。百年来，用混凝土以外的材料来建造建筑的技术因被遗忘而持续衰退。比技术缺失更糟糕的是，因为遗忘了技术，人们也失去了自己曾经建立起来的最重要的东西。20世纪的人类就像认知障碍患者一样，失去了与材料和技术相关的记忆。建筑师眼里只有混凝土，对其他材料熟视无睹。

混凝土最大的问题是它鼓励偷懒。如果用混凝土来做建筑，那么怎么组装，如何连接这些问题都不用考虑，只要确定了建筑的外形，接下来只需要按照外形做框架，然后把混凝土灰浆灌进去就可以了。所以刚出校门的学生也能做出混凝土建筑的设计图，只要画个外形，标注一下混凝土，看上去就像是设计图了。

雷姆·库哈斯曾把混凝土建筑比喻为大洪水中的"挪亚方舟"，而我认为，混凝土其实就是大洪水本身。大洪水将人类建立起来一切都冲走了，混凝土也冲走了建筑中所有珍贵的东西。我和我的团队也许就像挪亚，我们试图建造方舟来抵抗混凝土的洪水。混凝土跟洪水一样，是一种单纯、原始、有破坏性的建筑手法。黏糊糊的流质过些时间就会凝固起来，变成一具庞大僵硬的尸体。这时候后悔也来不及了。

把树枝编织起来，把石头、砖块堆积起来，这些施工方法都很费脑筋，不得不好好想、反复想。但这本身就是令人愉快的过程。

由于混凝土像大洪水般泛滥，20世纪的建筑比以往任何时代都乏味贫弱。让建筑再次变成愉快有意义的工作，正是我们设计的宗旨。一旦抛开混凝土，那么我们就不得不去思考，无法再偷懒了。

三个层面

当然，光思考是不够的，还需要灵感的启发。没有灵感，面对一种材料，可能也无从下手。因此，我想把30年来积累起来的灵感一次性地公开，从材料、方法、几何学三个层面来进行阐述。

第一个层面是材料的多样性。首先要认识到材料是各种各样的，选择要尽可能宽泛。一旦体验到了材料多样性的魅力，接下来视野就自动打开了。

第二个层面是材料的使用方法。思考材料要怎样去用，怎样堆叠，怎样借力支撑，怎样组合……各种方法的界线其实是模糊的，一边支撑一边组合，一边组合一边堆叠，这些情形都会出现，因为主角是人而不是空间，所以当然会有模糊的地方。我的目的不是为方法分类，而是告诉大家方法有很多，把方法这个工具递出去，给大家提供一些实践的灵感，鼓励大家使用这些工具，切入材料。

第三个层面是几何学。对材料进行运用，得到的结果会具有某种几何形态。有时候一种形态看起来杂乱无章，其实它也是一种几何形态。动物筑巢，也会不知不觉得到一个几何形态——没有比动物的巢穴更美的几何形态了。人类也是动物，想要做个巢，无意识地与材料纠葛，结果得到一个几何形态的东西，这种事情是经常发生的。

我认为，建筑之所以能打动人心，不是因为外观的几何形态，而是因为内在的几何学。20世纪的建筑只考虑外观的几何形状，按照外部的几何框架灌入黏糊糊的混凝土灰浆，而内在的几何学是欠缺的，因此乏味的建筑很多。我提倡的是与这种注重外观的几何体完全相反的建筑，即拥有内在几何学的建筑，一种生物学意义上的几何学建筑。

材料挑战的树形图

除了从材料、方法、几何学三个层面来揭示

我这30年来积累的灵感，我还做了一个树形图，希望能给今后要与材料打交道的人提供更多的启发。这个树形图展示了我如何挑战材料、如何坚持挑战的历程。首先，我想强调的最重要的一点是，与材料打交道需要花费很多时间。材料是一种具有自然属性、有生命的东西，当我带着某种方法、某种几何学的思考去挑战一种材料的时候，它就会有各种各样的抵抗，所以总是会出现各种不可预测的问题。即使在工期和预算内完成了一个项目，但总是会有问题留下，让你想再挑战一次。找到机会，我会在上一次经验的基础上，以新的技术再次挑战。然而这次又会发现新的问题。材料就这样持续激发着我的斗志。

这种挑战没有止境。在我的建筑生涯中，我想留下的不是建筑作品，而是一个实验性的研究室，一个与各种研究者和技术工作者一起研究材料，进行各种尝试的研究室。建筑设计只是研究室活动的一部分。

重要的不是个别的建筑作品，而是持续的努力，坚持不懈。真正的财富是坚持这件事本身。坚持的结果就是这样一个树形图。这个树形图不是一个简单的树结构，里面有缠连，有往复。从这个意义上来说，它不是树形，而是建筑师克里斯托弗·亚历山大所说的半网络结构（semi-lattice）。我经常遇到的情况是，做一个木材的建筑，结果需要借助石材或纤维；想要用简单的几何网格来解决问题，结果做出一个螺旋的几何形态。

最近，我开始思考什么才是我真正的作品——不是一座座单独的建筑，而是这个树形图及其背后的持续努力，是将这些建筑联系起来的理念。如果这个理念足够完善，那么即使我不在了，挑战还会继续下去。

20世纪的建筑师喜欢将建筑称为"作品"，就像艺术家把创作的有附加价值的商品称为"作品"一样。人们认为作品是一种与环境分离的特殊的东西，这种分离会赋予作品光环，使作品售出高价。而这种分离的作品破坏了20世纪的环境，现在还在增殖。

我认为，建筑完全没有必要成为一件"作品"，完全没必要与环境分离，也不需要特别的光环，不需要卖出高价。想卖出高价，建筑师就不得不做出一些勉强的东西，建筑就会变得扭曲——被"作品"的观念扭曲了。

建筑的价值不在于让每座建筑变得特别，其真正的价值来自不断失败，又不断从失败中学习，如此循环往复。从失败中吸取教训，下次争取做得更好，这样的信念才是最重要的，这个坚持的过程才是无价的。通过与材料的接触，我收获了最重要的东西。

树形图
Tree Diagram

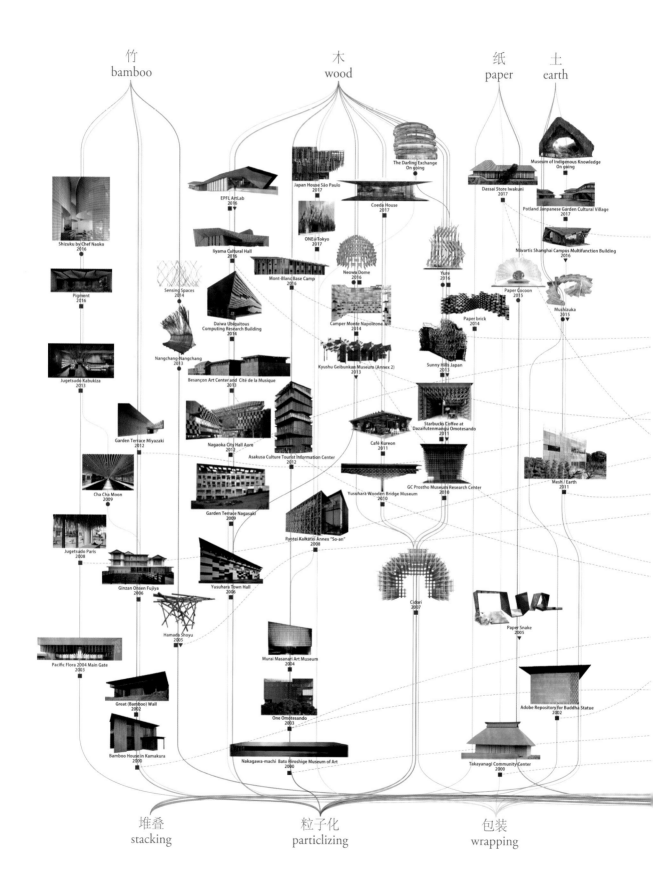

竹
bamboo

木
wood

纸
paper

土
earth

Shizuku by Chef Naoko
2016

EPFL ArtLab
2016

Japan House São Paulo
2017

The Darling Exchange
On going

Museum of Indigenous Knowledge
On going

Dassai Store Iwakuni
2017

Coeda House
2017

Pigment
2016

Iiyama Cultural Hall
2016

ONE@Tokyo
2017

Potland Janpanese Garden Cultural Village
2017

Sensing Spaces
2014

Mont-Blanc Base Camp
2016

Neowa Dome
2016

Yure
2016

Paper Cocoon
2015

Novartis Shanghai Campus Multifunction Building
2016

Jugetsudo Kabukiza
2013

Daiwa Ubiquitous
Computing Research Building
2014

Camper Monte Napoleone
2014

Paper brick
2014

Mushizuka
2015

Nangchang-Nangchang
2013

Besançon Art Center and Cité de la Musique
2013

Kyushu Geibunkan Museum (Annex 2)
2013

Sunny Hills Japan
2013

Garden Terrace Miyazaki
2012

Nagaoka City Hall Aore
2012

Asakusa Culture Tourist Information Center
2012

Starbucks Coffee at
Dazaifutenmangu Omotesando
2011

Café Kureon
2011

GC Prostho Museum Research Center
2010

Mesh / Earth
2011

Cha Chi Moon
2009

Garden Terrace Nagasaki
2009

Yusuhara Wooden Bridge Museum
2010

Ryotei Kaikatei Annex "So-an"
2008

Jugetsudo Paris
2008

Ginzan Ohsen Fujiya
2006

Yusuhara Town Hall
2006

Cidori
2007

Hamada Shoyu
2005

Paper Snake
2005

Pacific Flora 2004 Main Gate
2003

Murai Masanari Art Museum
2004

Great (Bamboo) Wall
2002

One Omotesando
2003

Adobe Repository for Buddha Statue
2002

Bamboo House in Kamakura
2000

Nakagawa-machi Bato Hiroshige Museum of Art
2000

Takayanagi Community Center
2000

堆叠
stacking

粒子化
particlizing

包装
wrapping

树形图的上方是材料分类，下方是方法，作品图片下方是几何学图示。
每件作品的描述和说明都由隈研吾本人执笔。

Each page includes a matrix of material, method, geometry.
All descriptions and captions in the book were written by Kengo Kuma.

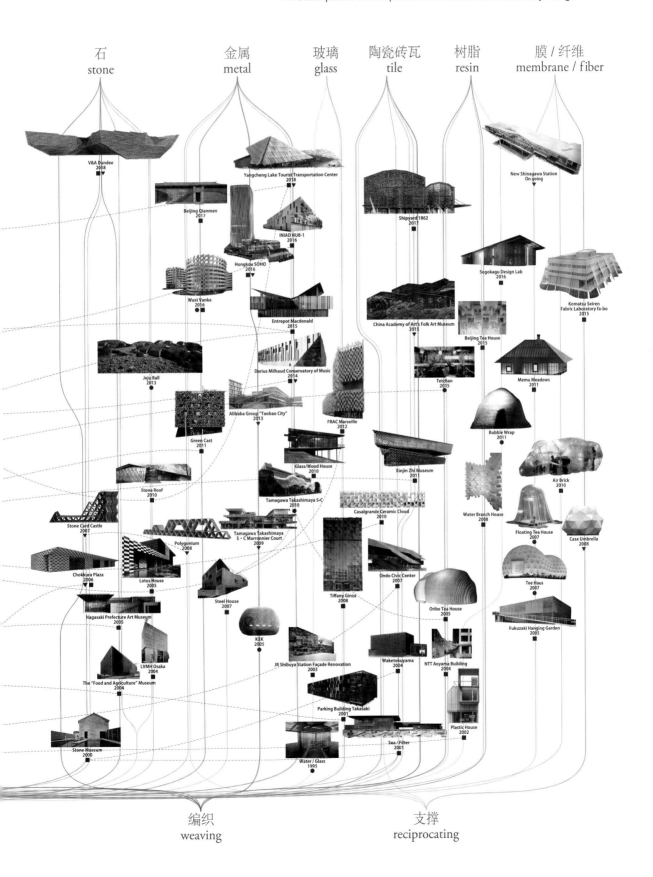

石
stone

金属
metal

玻璃
glass

陶瓷砖瓦
tile

树脂
resin

膜 / 纤维
membrane / fiber

V&A Dundee
2018

Yangcheng Lake Tourist Transportation Center
2018

New Shinagawa Station
On going

Beijing Qianmen
2017

INIAD HUB-1
2016

Shipyard 1862
2017

Hongkou SOHO
2016

Sogokagu Design Lab
2016

Wuxi Vanke
2016

Komatsu Seiren
Fabric Laboratory fa-bo
2015

Entrepot Macdonald
2015

China Academy of Art's Folk Art Museum
2015

Beijing Tea House
2015

Darius Milhaud Conservatory of Music
2014

Memu Meadows
2011

Jeju Ball
2013

Tetchan
2015

Alibaba Group "Taobao City"
2013

FRAC Marseille
2012

Bubble Wrap
2011

Green Cast
2011

Xinjin Zhi Museum
2011

Air Brick
2010

Stone Roof
2010

Glass/Wood House
2010

Water Branch House
2008

Floating Tea House
2007

Stone Card Castle
2007

Tamagawa Takashimaya S-C
2010

Casalgrande Ceramic Cloud
2010

Casa Umbrella
2008

Polygonium
2008

Tamagawa Takashimaya
S・C Marronnier Court
2009

Chokkura Plaza
2006

Lotus House
2005

Ondo Civic Center
2007

Tee Haus
2007

Steel House
2007

Tiffany Ginza
2008

Oribe Tea House
2005

Nagasaki Prefecture Art Museum
2005

KXK
2005

Fukuzaki Hanging Garden
2005

LVMH Osaka
2004

JR Shibuya Station Façade Renovation
2003

Waketokuyama
2004

NTT Aoyama Building
2004

The "Food and Agriculture" Museum
2004

Parking Building Takasaki
2001

Plastic House
2002

Stone Museum
2000

Sea / Filter
2001

Water / Glass
1995

编织
weaving

支撑
reciprocating

我认为建筑不是一件件单独的作品，而是一种持续性的工作。

一座建筑完成后，我们会有成就感，但也会有反省，会看到下一次需要解决的问题。当然，下一次又会有新的问题。这个过程会反复持续下去，我们每次都必须一步一个台阶地解决问题。

我们把工作的内容从材料、方法、几何学三个层面进行划分，然而有时想法也会从一个层面跳跃到其他层面中去。持续是一种力量，跳跃也是一种力量。

为了更好地展现我们的材料研究工作，我们把材料归纳为10种（竹、木、纸、土、石、金属、玻璃、陶瓷砖瓦、树脂、膜／纤维），方法归纳为5种（粒子化、编织、支撑、包装、堆叠），几何学归纳为3种（■网格、▼多边形、●螺旋）。

材料与方法以色彩不同的线相连，作品按时间顺序排列，材料与其他类别之间的跳跃用虚线连接。图中展示的连接只是一部分，实际上还存在更复杂的关联。

当代建筑背后的秘密

隈研吾：通过材料研究使当代建筑
重焕魅力

Backstory

Some notes on Kengo Kuma's project to re-
enchant Contemporary

Architecture through Material Research

杰弗里·基普尼斯

建筑常被认为是最古老的艺术形式，在西方哲学中甚至被称为"艺术之母"。然而作为一种专业划分的当代建筑，在现代的各种艺术形式中应该算是最年轻的。1963年美国全面完成了农村的电气化工程，意味着都市与乡村里所有建筑物的内部与外部都实现了电气化。这项国家战略的目的是建立以"个人自由"为基础的中产阶级社会。同时期，黎明期的空调、电视，以及室内室外24小时无处不在的照明等，成为第一代杀手级应用产品，与发达国家的现代交通、广播、飞机、汽车等既有技术相结合，进一步放大了技术所带来的日益增长的压力。当代建筑的诞生，我认为与这样的时代背景有着不可分割的关系。

到了20世纪90年代中期和后期，有线电视、卫星通信和电脑已经变得很普及，就像过去的炼金术演变为化学，占星术演变为天文学，一直在演变的传统建筑已经完全转化为当代建筑。与炼金术、占星术一样，传统建筑在前现代的全盛期，属于一种基于神秘主义、通灵式的手工业的原科学，在所有文化中都是为掌握着最大财富及权力的人群服务的。然而与炼金术转变为化学不同，传统建筑在向现代建筑转化的过程中，尽管经历过某种程度的尝试及自我认同，但并没有真正转变为一种纯粹的平等主义的科学。

不必理会历史学家的戏言，当代建筑并没有失败，反而可以说是成功了。它确实点明了标准化工业在哪些方面以及如何改善普通市民的生活，当然它对政治权力如何利用这些知识欠缺思考。最幼稚的是，它没有认识到，人们依赖的不全是科学和工程学上的知识，而是那些更加微妙、亲密、神奇的力量，类似杰里米·边沁所主张的"为最多的人谋取最大的幸福"。人们需要依赖这种力量去实现莱布尼茨所说的"一种哲学性的实践能力"，即能够在不同的地方为不同的人带来一种完全不同的价值体系——拥有最大限度的多样性和连贯性。结果，标准化的国际风格成了资本主义的标志，也成了霸权的表达途径，压制了建筑在表达文脉和文化多样性方面的深厚天赋——不论是在历史范畴还是全新的建筑语言方面，也不论是在形态、构成还是材料方面。这里，我称之为当代建筑的，正是近来那些天赋异禀者及天才的实践。

在20世纪这相当短的时间里，当代建筑在多种社会中都有存在，并出现分歧。建筑涉及的诸多专业技能被分割成多个领域，尽管它们互相交流，但至今仍大致保持着独立性。在建筑施工和工程学领域，主要是技术上的操作，以能否解决量产的问题作为衡量成功与否的标准。建筑形态的设计及与业主签约，主要是商业行为，以满足业主的需求及获得报酬为目的。而至于当代建筑，它与任何当代艺术一样，是一种文化性、艺术性的事业，其成就需要依靠批评家、策展人、收藏家、赞助人、有鉴赏水平的观众、特定客户（民间的或政府的），以及各种机构做出的阐释和评价来衡量。这些人（机构）往往会从更广阔的视野去评价建筑行为和建筑的专业价值。这恐怕是为了附加值，因为所有的建筑必须经过年度成本效益分析，同时满足安全、实用、舒适等基本标准，且如今还必须应对资源管理、环境保护等问题，以及其他社会、政治方面的问题。

从20世纪90年代后半期至今的20多年时间里，CAAD（计算机辅助建筑设计）与CAM（计算机辅助制造）等数字设计技术对建筑领域及行业产生广泛影响，导致了学术界乃至建筑业界的形态分化，如同5.4亿年前的寒武纪生命大爆发，地球上的生物门类出现了史无前例的爆炸式增长。同样的道理，人们很容易将丰富的资源和新的造型方法，以及主体结构材料的新型施工技术（令人想起一些可怕的事情吧？），与爆炸式的变化联系起来。持上述观点的读者，应该切实认识到了这样的状况。然而尽管如此，这些情况及其对当代建筑产生的影响，在同一时期究竟发生了怎样戏剧性且根本性的变化呢？对此，很多人恐怕只有一些非常模糊的认识。媒体技术极速发展，虚拟现实（virtuality）即便不是现实（actuality），但作为客观存在的一个状态，已流行并真正融入了人们的日常生活。那么，世界经济与政治的沉浮是个怎样的状态呢？看看吧：迪克·切尼？莎朗·艾普？迪奥·布兰度？某国间谍引发核攻击！酒吞童子、玉藻前、崇德天皇等魑魅魍魉的飞扬跋扈……对我来说，这一切看起来都非常无聊。

我感兴趣的是，以上述情况为背景，对20世纪60年代以来日本建筑的具体案例进行研究，探讨其对建筑史尤其是当代建筑史的重要性。之所以要针对这个情况做一个简单的叙述，是因为我相信这是必要的，而且今后还将继续就这个情况进行考察。

在1965年之前的当代建筑领域里，日本并非模仿者，甚至是资本主义国家中不同寻常的一员。这要归功于其历史、文化、地缘政治、地理、知识、技术、哲学、经济……若要一一列举的话，恐怕会长达数页（坦率地说，要是对各个项目进行详细说明的话，那更是需要长篇大论了）。所有这些出人意料且不协调的力量集中在一个点上，这真是令人意想不到的奇迹。不过我发现，迄今为止，几乎没有哪位学者对这一影响深远的情况做过深入的研究。其实，令人惊叹的证据随处可见。有七八代实践者，他们拥有杰出的才华、智慧和创造力，把日本传统文化的价值观与欧洲（德国、法国、荷兰、瑞士）的形式主义和哲学思想（如德国的存在主义与法国的现象学）这些相互矛盾的力量，以及美国的个人主义，完美地调和在了一起。这种调和为的是那些不属于以上任何一种文化的人或场所，这完全是一种独创，令人难以置信。

我在1990年就强烈地意识到了这个事实。然而对于自己的所见及感受，我自认并没有相应的批判能力和才智来进行说明，加上人们对这些事鲜有关注，我也不由得感到沮丧，便一直远远地关注。日本只有少数几个建筑师获得普利兹克奖，这与他们在文化领域和学术领域做出的深入研究相比，简直是微不足道的回报。在那段时期，我有幸结识了许多有才华的建筑师，被矶崎新、安藤忠雄、伊东丰雄、妹岛和世，以及许多与他们同样优秀人物所吸引。逐渐地，我对一些处于外围的建筑师（至少在我看来）产生了特别的兴趣。这些建筑师因为各种各样的原因，显得过于古怪、太过超前、愤世嫉俗。我不知道为何对他们感兴趣，也不知道自己对他们的评价是否客观，正如我前面所说，我只是远远地关注。这些人包括高松伸、长谷川逸子、北川原温、坂茂，等等。其中最让我感兴趣的便是隈研吾，当时他为马自达汽车公司在东京设计的M2项目（1991）刚刚开业，我在东京遇到了他。

我始终密切关注着他的作品，尽管我非常确定他本人对此一无所知。随着时间的推移，我逐渐认识到，他是一个不安分、极具天赋、聪明且有思想的建筑师。当时，日本国内及其建筑文化与大受欢迎的当代建筑进行着抗争，他也跟自己内在的矛盾冲突进行着抗争。到了2008年，隈研吾已经成为我研究日本建筑的主要课题，那一年他的著作《反造型》推出英文版（原书出版于2000

年）。在书中，他袒露了那些我猜测过的内在冲突。就这样，秘密被揭开了，这对我而言是突然降临的真相，对他则是一个胜利。

在《反造型》中，隈研吾以一种深思熟虑的论证方式，阐述了对当时建筑状况的批判。当时根深蒂固的建筑思想是将建筑物视为独立的造型体，与土地乃至人们生活的环境是割裂的。他论述道，这些造型体无论是智慧的产物还是审美的产物，都体现着对建筑与周边环境之间关系的否定，而这种关系对建筑来说是非常重要的；建筑随着时间的推移与周边的环境一同成长进化，需要一种合作互助的健康关系。

《反造型》的强大之处在于，它为当代建筑的亲密性这一崭新的研究课题提供了一个框架。这里所谓的当代建筑指的是这样一些建筑：它们愿意并能够全盘接受媒体饱和的现实，以及高度散漫的建筑环境。它们曾受惠于20世纪70年代至90年代的那些理论，包括批判性、抵抗、文脉主义，尤其是两大现象学理论——场所精神（场地）与认知（经验）理论。这些理论曾经有效引导了当代建筑的创造和激情，而现在已经不再起作用。

过去的30年里，伴随着建筑形态的爆炸式发展，各种新的物质材料也层出不穷，比如屏幕、照明和背光，具有反射（折射）功能、经久耐用的先进涂层、色彩饱和的高分子材料，等等。这些戏剧性、刺激性的物质材料能够引起人们的短暂注意，以抵抗发

达国家的两种空间特征及其对地缘政治经济的潜在渗透，尽管有些地方的基础设施甚至尚未完善。如果去到那些尚未接通自来水的非洲村落，你会发现大多数家庭都有卫星电视、电气设备，到处都有智能手机和索尼PlayStation 3游戏机。我们可以看到两种情况：第一，信息超载对我们的注意力形成了持续的冲击，使得建筑对注意力应有的控制被削弱。注意力的各种状态——散漫的、随意的，或细致的、集中的，有时甚至是精读式的——会在瞬间发生非线性的跳跃。第二，正如已经论述过的，有一种不安定的存在论并不关心"物性"——卡通人物、照片、建筑物、音乐样片、推文，以及电影中的一个场景，乃至事实，它们都是可互换的，也几乎无法区分；真实、现实、事实、虚构，这些随着语境或视角的变化，成为流动的、可变的东西。这对于建筑来说，会产生一些特别且有趣的问题，也会带来新的可能性。只有那些为过去所困的傻瓜才会称之为危机。

隈研吾材料研究的巧妙之处在于，他一直在探索如何用传统和当代的物质材料，去激发当代社会对建筑的全方位关注。他没有怀旧，也没有指望噪声和竞争就此消失。从冷静思考到煽情诱人，从瞬间到永恒的时间跨度，从空间到空间，从物质到物质，他只是希望你在这个过程中去发现。

那些在混凝土的凹陷中摇晃的影子，难道真的只是自然光投下的普通阴影吗？卡恩诗意地称之为"被消耗的光"，真的是这样吗？还是为了更

醒目、更活泼而涂上去的？这当然是玩笑话。但是请记住，所有的建筑师都是魔术师，都是出色的"骗子"。正如你将要发现的，隈研吾和他的团队能够开发出一种新的木材组合方式，具有比闪烁的霓虹灯还要持久的魅力和动感。但他真正的魔术是，在你还没有注意到发生了什么的时候，在你还没有意识到自己已经进入了一个来自过去的传统空间的时候，你的情绪已经在不知不觉中舒缓下来，被静寂和深沉淹没。这是他最擅长的魔术，每次都有惊人的效果。我不会告诉你们他是如何做到的（因为我至今也没有搞懂），但是身处那个展览空间（"隈研吾的材料研究室"展），你或许能找到答案。

杰弗里·基普尼斯
建筑评论家、理论家，现为美国南加利福尼亚建筑学院客座教授。

粒子化：
粒子的艺术与科学

Particlized:
The New Arts and Sciences of Particles

马里奥·卡珀

当年，隈研吾作为一位颇受争议的后现代古典主义建筑师首次获得了国际声誉。他在20世纪90年代的著作中一直坚持的反现代主义立场，后来却发生了意外的转向。他曾把枪口对准建筑形态，后来却转向了现代主义标榜的产业技术的逻辑。现代主义的目标是规模经济，即大规模生产标准化的建筑构件。众所周知，隈研吾很快就表露了对混凝土的厌恶（如果他讨厌的是钢铁倒也不奇怪）。他对混凝土这种方法进行了毫不留情的抨击，特别是现浇混凝土（假装）具有的结构连续性，以及周围的建筑师们为了追求圆滑的面或体块而努力学习的最新计算机设计工具。[1]

2000年，他首次以图书的形式[2]发表了开创性的理论"反造型"，不久又发展出了所谓的"粒子化"建筑风格——由松散的粒子形成模糊的轮廓，并很快成为他的标志和口号。按照当时他的说法，这种松散的粒子状态不仅仅意味着某种建筑形态，还需要从更高的维度去理解："那是一种世界观，一种哲学……过去，这样一种平面的（散碎的）状态只可能被认为是一团混乱，无从下手，令人不快……可是，以计算机为代表的现代科技可以在不进行分类整理、结构性整合的情况下，对一团具体的粒子进行处理。为此，需要事先把每个元素从连接或结构中解脱出来，置于自由的状态。这样，具体的东西就能保持着鲜明的具体性，与世界直接连接起来。这是我想象中松散的粒子化世界的景象，也是我认为的自由景象。"[3]简单来说，粒子化与所有现代的科学原理及技术基础都是背道而驰的。

现代科学与技术，也就是我们在学校中学习的科学，以及推动工业发展的技术，旨在让事物变得简单。我们面临的世界是一片毫无意义的混乱，为了对它进行了解、预测并采取行动，我们需要将其转化为简单的公式或规律，这样才能完全纳入我们人类的大脑。然而我们的大脑无法接收过于琐碎的信息，因为记忆的容量是有限的，处理大量的信息也很花时间。所以，科学倾向于采用数学方程式、函数等简单便捷的工具，依据极有限的可测量因素去确定事物的因果关系。以类似方式建立的静力学、材料力学正是现代科学的重要成果。19世纪，弹性力学用当时最强大的数学工具——微积分——来描述建材应力变形的理想状态，即在各个尺度层面，完全各向同性（isotropic）、均质，保持着连续性。莱布尼茨和牛顿的微积分是古典数学的顶点，而数学用无限大和无限小这种现实中不存在的理想概念去模拟自然。虽然微分学是记叙无穷小量的，然而用这个方法最多只能把自然现象描述为一条平滑、连续的曲线（表现能够函数化的平滑、连续的变化）。这并不奇怪，因为莱布尼茨认为连续性是物质界的普遍法则。在被现代科学奉为金科玉律的莱布尼茨的古典世界观中，自然界是没有跳跃的。

然而不论是过去还是现在，天然的建材大致是不存在连续性的，它们往往充满瑕疵，有孔洞、裂缝、节瘤等。所以工业革命时期的科学家们努力开发人工材料，想让材料具有连续性、均质性，极力贴近数学函数。其中，工业用钢材从头到尾都具备各向同性的弹性，正是完全符合要求的绝佳材料。稍晚出现的钢筋混凝土则仅次于钢材。19世纪晚期至20世纪的土木工程学是建立在弹性力学的理论基础上的，然而弹性力学的局限性也是众所周知的。因为弹性力学的对象无论是物质还是数学，都是以绝对的连续性为前提的，比如这个理论也许可以计算出以钢材建成的埃菲尔铁塔的变形，但无法计算出围绕花园的3米高砖墙的耐久性。在莱布尼茨的设想中，墙砖之间的水泥填缝是不存在的。而这正是某种"自然的跳跃"，是莱布尼茨的数学所无法描述的。

20世纪中叶开始，"有限元分析"（FEA, finite element analysis）这种计算模式发展了起来，即把一个三维网格覆盖在解析对象上，将其分割成微小的粒子，再对各个微小粒子进行分析计算。不过，有限元模型会产生庞大的数据，直到最近，凭借高性能的计算机，人们才能真正对其加以利用。从进化论到社会科学，从热力学到材料科学，20世纪70年代开始相继出现的这些后现代科学理论——如今被统称为复杂性理论（complexity theory）——虽然属于不同领域，目的也不同，但都关注非连续体。这些理论的数学衍生品中，有一种"细胞自动机"（cellular automata）模型，通常被认为在模拟自然现象方面非常有效。它首先将连续体分割为细胞列，然后对细胞与相邻细胞之间相互作用的规则进行记叙。然而长期以来，人们认为这种数学上很有趣的粒子化手法没有实用价值。直到最近，计算机取代了人工计算，才让事情变得轻而易举。

可以说，这对今天所谓的数据科学是一个巨大的启发。与人类相比，计算机是如此强大快速，可以直接对现实进行处理；面对人类束手无策的复杂现实，计算机不必依赖数学捷径或科学解释就能处理。

人类不擅长处理复杂的东西，但计算机擅长。比如连续函数用数学来表达，通常就是y=f（x），很少的几个字母符号就表达了满足这个条件的所有的点（比如从图形上来看，就是同一曲线上所有的点）。可是计算机不一样，它会将这个公式产生的一个个的点转换为数量庞大的坐标值，生成一个无限长的列表。这个列表对我们人类来说没有任何意义，不要说使用，仅仅是编辑就需要漫长的时间。然而计算机并不需要理解这个世界，它工作起来比我们快得多，在我们试图算出一个优雅的数学公式的时候，它早已迅速正确地处理好了那些离散的数据。或者可以想象一下电话簿，要想在有着上百万个名字的电话簿中找出需要的

那一个，我们首先要把名字按照字母顺序排列起来——不走这一步（一种捷径），我们是找不出来的，一行一行地找太花时间了。可是计算机却擅长做这类事，不论名字是按怎样的顺序排列的，或者根本没按顺序排，计算机就这样一个一个地读下去，而且读得比我们快得多。计算机最擅长信息搜索，所以不用事先对数据进行分类排序。有了这样强大的搜索能力，从前那种按照个人、职业分门别类的电话簿现在已经没人用了吧，很多国家甚至已经停止发行电话簿了。

这种例子还有很多很多。总之，这种知识的变化已经给包括日常应用技术在内的各个领域带来了显著的影响，计算机科学家们在拼命寻找新的术语来描述这场前所未有的技术革命。数年前大家都在说大数据，如今热门的话题则是人工智能（AI）。当然，设计师们也已经注意到了。看看数字革新者们正试图从这空前的数据世界中创造出什么——设计师们凭借大数据或人工智能，极力提升形态及结构上的"粒度"（granularity），试着营造出设计感。有时人工智能能够做出以假乱真的超高解析度，比如一个由40亿个三维像素填满的面，却没有人能够像计算机那样一个个地数出40亿个三维像素。究其实质，这显然并非人类逻辑的产物。再比如，用机器人装配标准或非标准的零件，如果没有先进的人工智能工具，连数据管理都做不到。[4]总之，人工智能的内部结构与

我们人类从古至今鲜有变化的大脑是两种完全不同的运作方式，后现代科学的手法利用计算机获得了名为"离散化"（或者粒子化）的新形式，并以肉眼可见的方式具体呈现出来。

　　隈研吾最近的作品，似乎神秘地弥合了传统工匠不可量化的直觉与电子机器的超人类逻辑之间的隔阂，那是因为今天的人工智能正是这样：最先进的计算机以"后科学级的速度"去驱动非数理思维的"前科学逻辑"。标准化反映的是工业革命的逻辑，粒子化反映的则是数字革命的逻辑。研究人工智能的科学家们已经开始接受这样的世界观，他们对那些既是技术文化也是意识形态的粒子化实践一定非常赞同，而隈研吾早在20年前就独自开启了这一历程。

1. 隈研吾：《材料·结构的细节》，彰国社，2003 年，第 7 页。

2. 隈研吾：《反造型——溶解·击碎建筑》，筑摩书房，2000 年。

3. 隈研吾：《材料·结构的细节》，第 15 页。

4. 关于超高解析度，参考迈克尔·汉斯梅尔（Michael Hansmeyer）、本杰明·迪伦布尔（Benjamin Dillenburger）或"杨 & 阿雅塔"（Young & Ayata）近期的作品。关于施工机器人及"局部与整体"（mereological）的构成，参考伦敦大学巴特莱特建筑学院的吉勒斯·雷齐（Gilles Retsin）、曼纽尔·希门尼斯·加西亚（Manuel Jimenez Garcia）、丹尼尔·科勒（Daniel Kohler），以及南加利福尼亚大学（洛杉矶）建筑系的乔斯·桑切斯（Jose Sanchez），或苏黎士联邦理工学院（ETH）的格拉马齐奥（Gramazio）、科勒（Kohler）等人的作品。

马里奥·卡珀
伦敦大学巴特莱特建筑学院雷纳·班纳姆建筑史和建筑理论教授，重点研究建筑理论、文化史、媒体与信息技术的历史关系。著作《机械复制时代的建筑》（*Architecture in the Age of Printing*）被翻译成多种语言。其他著作有《第二次数字转变：超越智力的设计》（*The Second Digital Turn: Design Beyond Intelligence*）、《字母与算法》（*The Alphabet and the Algorithm*）、《建筑中的数字化转变：1992—2012》（*The Digital Turn in Architecture, 1992—2012*）等。

存在与物的融合

关于隈研吾建筑的考察

Connecting Beings And Things

Reflections on Kengo Kuma's Architecture

理查德·斯考菲亚

隈研吾的建筑最引人注目的一点在于，他能够把自己的介入，即"作为一位建筑师的存在"，表现得很明确，同时又显得若无其事。他不会直白地表现自己，却能够对所有行为给出"此时""此地"的精确定义。这种令人心生向往的建筑，平和沉稳地给自己赋予存在的光环。他对于那种为显示自己的存在而去牺牲使用者利益的建筑形态是非常抗拒的。

石头和金属以一种不可思议的关系结合在一起，层叠的杉木条造成光线从明到暗的流动，透光的格栅……隈研吾的建筑似乎向我们许诺了一个世界，在这个世界中，所有的对立都得以化解，而所有事物在多重合作的基础上自由发展、和谐共处。

对隈研吾的建筑进行解读，可以就以下三点进行考察：其一，他如何在各种完全不同的材料之间建立亲密关系，运用了哪些独创且细致的处理方式；其二，人如何适应周边环境的变化，在身体上做出相应的改变；其三，密切关注他对身体的深入思考（他的建筑仿佛是编织成的，就像一件合身的衣服）。

与材料合谋

隈研吾设计的建筑中，从未出现过任何戏剧性或暴力性的东西。他的建筑更像是用各种不同的材料精心组装成的某种可触摸的装置，展示着自己长期以来的种种精巧构思。

　　建筑的构件完全没有屈从于创作者的意志，因为建筑师的根本职责在于理解并爱护它们，尽一切努力保持它们原本的属性，最后协助它们完成使命。

　　如法国马赛当代艺术中心，设计的重点在于让捕捉到的光线成为建筑的主要构成元素。1,700枚涂釉的玻璃悬挂在外立面上，吊装固件被隐藏起来，整个建筑看起来就像覆盖着鳞片。射入的光线，使建筑仿佛被光轮环绕着一般，散发着超现实的光辉。一位独自生活在法国奥弗涅地区的玻璃工艺大师对这些"鳞片"进行了反复的实验，以确认透明度、质感、光的扩散性能及受热时的状态等。而建筑师则把全部精力放在了如何用好这些"鳞片"上。

　　日本栃木县高根泽町的巧坷垃广场又是另外一个故事。当地出产一种著名的白色石材——大谷石，曾被弗兰克·劳埃德·赖特用在原东京帝国饭店上。大谷石比较脆，设计师于是以钢铁为补强材，将其与石材编织在一起，做成了拥有菱形孔洞的透光的屏风墙。

　　离巧坷垃广场不远的那须芦野的石头美术馆，又让我们看到隈研吾其他的创意。在那里，石头被当成木材来使用，薄薄的石板条水平组合，以木工常用的榫卯结构与支柱连接在一起，这与古希腊建筑师模仿木结构建筑建造大理石神殿的手法有共通之处。

　　此外，位于日本石川县的"小松精练纤维研究所fa-bo"，本来是一座老旧的混凝土建筑。因建筑不符合日本现行的抗震标准，建筑师使用碳纤维的绳索对它进行了抗震加固。围绕建筑物一圈的绳索看上去就像"妈妈的裙子"。

　　就像这样，建筑师在各个不同的场所，从各种不同的材料开始思考，把各种材料结合在一起，让它们大展身手。这些材料共同合作、互相搭配，形成一座座迷人的建筑物。

来自本能的秩序

　　隈研吾建筑的另一个特征，是他对人类本能行为的关注。这些行为来自祖先，仿佛已经铭刻在人类基因中。不需要学习，鸟就会筑巢，狼就会做窝，蜘蛛就会结网。同样的，人类对材料也会做弯折、堆叠、组合等操作。这些行为以传统的形式延续下来，比如折纸、千鸟格。隈研吾的建筑有时就像逆流而上、追溯源头的河水，唤醒我们隐藏在内心深处的动物本能。做建筑的人与自己的建筑有机结合在一起，体现出一种来自本能行为的"简单关系"。

　　不需要用到钉子、锤子、螺丝刀、绳子这些东西，只需把事先做好缺口的木板条简单嵌合起来——隈研吾为法国巴黎国际当代艺术博览会（FIAC）在巴黎杜伊勒里花园做的木结构装置"晃动"就是这样一件作品。和他的很多建筑作品一样，搭建工作不需要有经验的工匠，谁都可以做。这件作品还让人联想到日本表参道上的店铺"微热山丘"，杉木条组合成复杂的形态，远远看上去就像远古人类的巢穴。

对于金属板，也不必粗暴地切割，而是用弯折的方式保持整体的完整性，比如多次弯折就可以得到一个房子的屋顶。推而广之，还可以用这种方法解决打开墙面的问题，也可以打破封闭的空间。比如普罗旺斯艾克斯的达律斯·米约音乐学院那史诗般的外观——外墙由铝板折叠而成，门窗隐藏在褶皱的阴影中，整座建筑看起来仿佛带有一种魔力。

很快，建筑师又在巴黎一个令人意想不到的地方——奥斯特里茨火车站，挑战了木材与金属的弯折，在轨道上方做了一个蝴蝶群般的屋顶。

还有用容器快速堆垒起来的临时避难所。似乎，所有的建筑物都与生存这个概念有着内在的联系。作为最低限度的生存所需，难民营可以说是所有住宅的基本原型。

同样有趣的是通过草图去观察那些隐藏在作品背后的东西：灰色铅笔勾勒出来的轮廓透着一种紧张感，还有密集的点、线及其他图案……设计师的笔触仿佛是在回应工匠们已经遗忘了的动作；那些冲动的纹理，像镜子一般反映着纠缠的结构。

身体与流动

对于隈研吾这些难以分类的建筑实践，需要着重考察的最后一点是他对人的身体及其延伸的思考。在日本传统建筑中，人的身体或坐或躺，动作很少，属于独立的元素。而20世纪60年代后期，在日本前卫的新陈代谢运动中，人的身体被视为一个整体，一个动态、有生命的集合体。这里所说的身体，并非勒·柯布西耶所说的标准化的人体，而是对光、热、噪声、寂静很敏感，容易受感情、情绪、好恶等影响的人体，是一种不稳定的存在。建筑师要触发它，同时也要保护它。

在那珂川町马头广重美术馆项目中，隈研吾并没有追求任何特殊效果。这座美术馆像一个长条形仓库，交叉的通道通向展厅，也通向美术馆背后树木繁茂的山丘。坚实的钢架结构上按一定间隔排列着温暖的杉木板条；墙面通过薄膜控制着光的摄入量，让参观者从阳光耀眼的室外进入展厅时，眼睛能够适应为保存、展示版画而调暗的光线。在贝桑松艺术文化中心，当代艺术作品的展示空间是用木材和玻璃组成的格子构成的，建筑本身已经消失，让参观者仿佛置身于草木茂盛的森林，光线透过些许枝叶落入林中。

仍在建造中的丹尼斯普莱耶尔火车站，需要应对每天约25万人的客流量。明亮的中庭中，人流沿着巴比伦通天塔般的斜坡道不断螺旋上升，通过轨道上方的架桥到达乘坐的列车所在的站台。

未完成的境地

　　隈研吾的建筑似乎拒绝成为一个完成的造型体或一种封闭的形态，因为未完成的作品能够释放出更多的可能性。那珂川町马头广重美术馆的形状就像一座常见的农用仓库，并没有特别显眼的地方。似乎，重要的不是它整体的形状，而是刹那间投射在地面上的木格栅的影子、衬托的游客身影，以及各种不同的空间特质……巧坷垃广场的几何形状也不重要，重要的是那些断断续续的肌理，就像一些散乱的乐章，让往来的行人产生共鸣。当古老米仓的镂空墙用在了新的商店建筑上，空间便产生了新的可能性，建筑也随之消失了……

理查德·斯考菲亚
1955 年，出生于法国尼斯。
1980 年，毕业于巴黎 UP 6［现巴黎 - 拉维莱特国立高等建筑设计学院（ENSA Paris-la-Villette）］，取得法国建筑设计师文凭（DPLG）。
1991 年，获得"年轻建筑师专辑奖"（The Album of the Young Architecture），并在巴黎创立理查德·斯考菲亚建筑事务所。
1992 年，任法国建筑学院教授。
2015 年，成为法国建筑学会委员。
2017 年，获"法国艺术与文学骑士勋章"。

竹

Bamboo

竹子是我的老朋友。我横滨老家的房子背后就有很大一片竹林。横滨的山野和人离得很近，在我出生成长的大仓山地区，沿着山脚有一排农舍，朝着稻田的第二家就是我的老家。

小时候我就在这样的山野中游玩，几乎每天都会穿过房子背后的竹林进山，爬上山脊。竹林中充盈着绿色的光线，就像另一个世界。我会用手牢牢地抓住竹竿，掂量着它的韧性，借力登上陡坡。上了山脊，天空很开阔，有风吹过。

竹子生来就是直的，这种强烈的直线性是其他树木所没有的。建筑师总是会被几何形态所吸引，因此很多建筑师都偏爱竹子这种材料。勒·柯布西耶的弟子坂仓准三喜欢用竹子设计家具，数寄屋建筑大师吉田五十八也为京都的料亭"冈崎鹤屋"做过竹椅子。

对京都桂离宫有着高度评价的德国建筑师布鲁诺·陶特，1933年赴日的次日就来到桂离宫，据说在入口前看到一片名叫"桂垣"的竹墙就痛哭起来。那种围墙现在看起来也很不可思议，是把竹林外围活的竹子直接弯折起来编成墙，这种墙介于自然物与人造物、野性与几何形态之间。这个"之间"的状态是由竹子的韧性和几何学创造出来的。

布鲁诺·陶特此后旅居日本三年，设计了两座住宅和一些物品。其中一座住宅就在我做的

Bamboo is an old friend, because there was a large bamboo grove behind the house where I grew up. I was born and raised in the Okurayama district of Yokohama, where the village hills were unexpectedly close. Our house was the second one in from the foot of one such hill, between a row of farmhouses and rice paddies.

I used to go and play at the farmhouses. Every day I would step into the bamboo grove behind them and climb to the top of the hill. It was a separate world inside the grove, bathed in green light. I would use my hands to climb the steep slope, grasping the trunks of the bamboos and measuring their resilience. When I reached the top, the sky would open up and a breeze would blow.

By its nature, bamboo is straight. It has a linearity not seen in other trees. This may be why it appeals to architects, who tend to be attracted to geometry. Le Corbusier's disciple Junzo Sakakura designed furniture from bamboo, and Isoya Yoshida, the master of the sukiya style, designed bamboos chairs for the Okazaki Tsuruya guest house in Kyoto.

The German architect Bruno Taut, who loved the Katsura Imperial Villa and visited it on the day after his arrival in Japan in 1933, is said to have been moved to tears by the sight of the katsuragaki bamboo fence at the entrance. This fence is still a remarkable sight, made up of living bamboo stems from the grove behind the fence, bent and woven into place. It is a quintessential intermediary between the natural and artificial, and between geometry and untrammeled propagation. It is an

1

建筑"水·玻璃"（1995）的旁边，叫作"日向邸"（1936），里面有竹子做的墙和灯具。

可是在20世纪，竹子很少用于建筑。原因是竹子干燥后容易开裂，不适合用作结构材料。

然而我们一直在挑战这个难题。在混凝土填充钢管技术的启发下，我们把竹子内部的横隔去掉，灌入混凝土做成柱子（镰仓的"竹之家"，2000）。

我们了解到南美洲有一种名叫瓜多竹（Guadua）的竹子，壁厚且不易开裂，就从哥伦比亚用集装箱运回来，将其用在一个酿造酱油的老厂房改造项目上（滨田酱油，2009）。这种竹子确实不易裂，但是组装衔接的工作却困难重重。竹子是一种精细、高冷的物种，但这也正是它的魅力所在。

intermediate space created by the resilience and geometry of bamboo.

During his three years in Japan, Taut designed a number of industrial products and two houses. One of the houses is the Hyuga Villa (1936), which stands near my Water / Glass project (1995). Bamboo is used in the villa for walls and light fixtures.

But the 20th century never used bamboo on a major scale to produce architecture. The reason is that bamboo splits when dried, making it difficult to work with as a structural material.

We have returned often to the challenge of overcoming this limitation. At Bamboo House in Kamakura (2000), we took a hint from concrete filled steel tubes (CFT) and removed the node walls from bamboo and filled them with concrete.

After learning that there is a thick-walled and split-resistant type of bamboo called Guadua that grows in South America, we had it shipped by container and used it in the renovation of an old soy sauce brewery (Hamada Shoyu, 2009). The bamboo itself does not split, but the joints were difficult. Bamboo remains a lofty and delicate material—not easily approached, but attractive for that very reason.

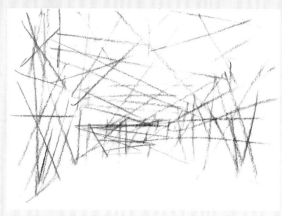

2

3

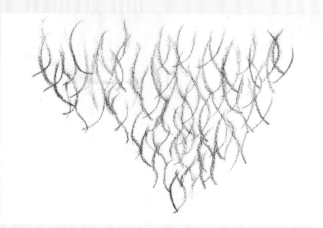

Sketches by Kengo Kuma
1. Great (Bamboo) Wall
2. Hamada Shoyu
3. Nangchang-Nangchang
4. Sensing Spaces

隈研吾手稿
1. 长城脚下的竹屋
2. 滨田酱油
3. 以心传心
4. 感知空间

4

长城脚下的竹屋
Great (Bamboo) Wall

位置：中国北京

设计时间：2000.12—2001.04

建造时间：2001.04—2002.04

主要用途：度假村式酒店

占地面积：1,931.57m²

总建筑面积：528.25m²

合作者：奥雅纳工程顾问公司（大中华地区）（ARUP China）、洛可可景观设计（Rocco Landscape Design）

结构顾问：中田结构设计事务所
（Nakata Structural Design Office）

建筑施工：北京市第三住宅建筑工程公司（Beijing Third Dwelling Architectural Engineering Company）

竹子因为容易开裂，通常只用作装饰材料。为了能把它用在结构上，我们去掉了竹子的横隔，填入钢筋混凝土，以混凝土填充竹筒替代混凝土填充钢管，实验性地用在了镰仓的"竹之家"中。

之后，我们在做中国长城脚下的这家别墅酒店的设计时，内部外部都用了竹子。直径60mm的竹子也按60mm间隔排列，整座建筑都控制在一个节奏上。为了延长竹子的寿命，我们用钢板加热竹子后，再给竹子上一层油。在日本通常不使用这种方法。

Traditionally, bamboo has been used only as a finishing material due to one fatal flaw: it is prone to splitting. To use bamboo structurally, we created CFB (concrete-filled bamboo) tubes by removing the nodes and inserting steel and concrete. In our design for Bamboo House in Kamakura, we experimented with these tubes as an alternative to CFT (concrete-filled steel tube) .

Next, in our design for this hotel beside the Great Wall of China, we used bamboo exhaustively on both the interior and exterior. We arranged bamboo members, 60 mm in diameter, at intervals of 60 mm, based on komagaeshi (a traditional detail where members of a given width are lined up at intervals of the same width). The detail enabled us to bring the entire design under the sway of a single rhythm. To ensure the bamboo would have a long service life, it was heated on steel plates and coated with oil. Ordinarily, this method is not used in Japan.

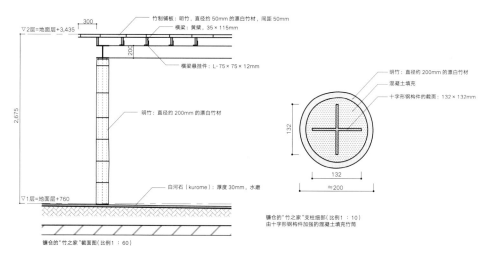

▽2层=地面层+3,435

300

200

竹制铺板：明竹，直径约 50mm 的漂白竹材，间距 50mm

横梁：黄檗，35 × 115mm

横梁悬挂件：L-75 × 75 × 12mm

2,675

明竹：直径约 200mm 的漂白竹材

明竹：直径约200mm 的漂白竹材

混凝土填充

十字形钢构件的截面：132 × 132mm

132

132

≒200

▽1层=地面层+760

白河石（kurome）：厚度 30mm，水磨

镰仓的"竹之家"支柱细部（比例1：10）
由十字形钢构件加强的混凝土填充竹筒

镰仓的"竹之家"截面图（比例1：60）

镰仓的"竹之家"，选用直径 200mm 的竹竿，先将横隔去掉，再将 132 × 132mm 的十字形钢构件插入竹竿中，最后浇入混凝土，成为竹版的混凝土填充钢管，即混凝土填充竹筒。

Bamboo House in Kamakura. Cruciform steel members measuring 132 ×132 mm were inserted into giant bamboo pieces with a diameter of 200 mm from which the knots were removed, and concrete was poured in to create bamboo CFT, or in other words, "CFB".

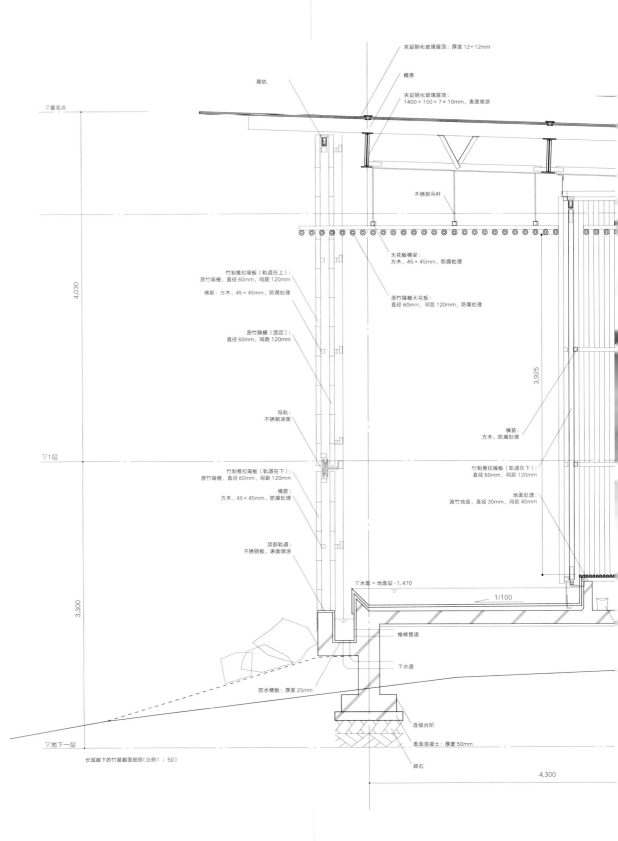

夹层钢化玻璃屋顶：厚度 12+12mm

悬轨

橹条

夹层钢化玻璃屋顶：
1400×150×7×10mm，表面喷涂

▽最高点

不锈钢吊杆

天花板横梁：
方木，45×45mm，防腐处理

竹制推拉隔板（轨道在上）：
原竹隔栅，直径 60mm，间距 120mm

横筋：方木，45×45mm，防腐处理

原竹隔栅天花板：
直径 60mm，间距 120mm，防腐处理

4,030

原竹隔栅（固定）：
直径 60mm，间距 120mm

3,925

导轨：
不锈钢涂黑

横筋：
方木，防腐处理

▽1层

竹制推拉隔板（轨道在下）：
原竹隔栅，直径 60mm，间距 120mm

竹制推拉隔板（轨道在下）：
直径 60mm，间距 120mm

横筋：
方木，45×45mm，防腐处理

地面处理：
原竹地面，直径 30mm，间距 40mm

顶部轨道：
不锈钢板，表面喷涂

▽水面 = 地面层 -1,470

1/100

3,300

维修管道

下水道

防水檐板：厚度 25mm

连续台阶

▽地下一层

长城脚下的竹屋截面细部（比例 1：50）

盖面混凝土：厚度 50mm

碎石

4,300

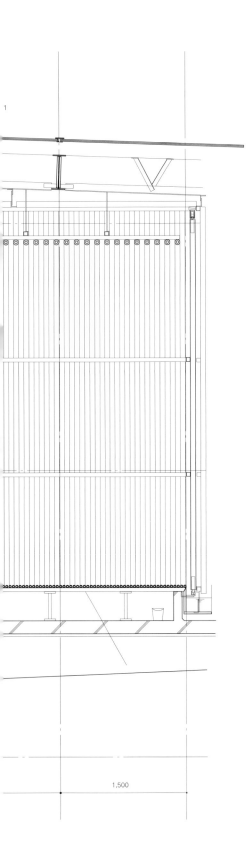

1

1,500

粒子化
particlizing

包装
wrapping

网格
grid

滨名湖花博会主入口
Pacific Flora 2004 Main Gate

位置：日本静冈县滨松市
设计时间：2001.04—2003.03
建造时间：2003.06—2003.12
主要用途：大门
占地面积：211,611m²
总建筑面积：2,445m²
结构顾问：TIS & Partners株式会社（TIS & Partners）
建筑施工：株式会社水野组（Mizunogumi）

我想做一个能给入场者带来某种特殊体验的大门。无数直径60mm的竹子从大门上部的结构体上悬挂下来，入场的游客仿佛穿过一片人工竹林。为了防止竹子开裂落下，所有竹子都打了螺栓，再用绳子绑住。在后续的项目中，我们反复使用过这种从顶棚垂挂线形材料的方式，而且用过木材、金属等多种材料。

I wanted to create an exhibition gate that would give all visitors a special experience—something akin to passing beneath a man-made bamboo forest. To achieve this, we suspended countless pieces of bamboo, 60 mm in diameter, from a gate superstructure. All pieces were penetrated with bolts and bound with strings to prevent them from falling, even if they split. In subsequent projects, we reused this detail for hanging countless linear members from a ceiling, substituting the bamboo with other materials such as wood and metal.

粒子化
particlizing

网格
grid

滨田酱油
Hamada Shoyu

位置: 日本熊本市
时间: 2009
主要用途: 商店
占地面积: 2,046.06m²
总建筑面积: 1,535.52m²
结构顾问: 江尻建筑结构设计事务所
（Ejiri Structural Engineers）
建筑施工: Cleanspace

我们用一种干燥后不易开裂的特殊的竹子——南美洲的瓜多竹，来挑战竹结构。编织竹篮等生活用品的技法中有一种不规则的"乱编法"，基于此，我们构思了一种"乱组法"：将竹子连接起来，灵活地支撑起老厂房的零碎结构，成为一种介于装饰和结构之间的存在。最大的困难是竹子的连接工作，我们的对策是在连接部分注入环氧树脂，再用螺栓固定。

To meet the challenge of realizing a bamboo structure, we used a special type of South American bamboo Guadua, that will not split, even when dried. A design and structure with an intermediate presence was created while flexibly responding to the patchwork of fragmented structures in old storehouses that consist of a "Yatara random braid" based on the "Yatara random weave" used in bamboo baskets and other items. The biggest obstacle in bamboo structures is the joints. Here, we injected joint portions with epoxy resin, then fastened them together with bolts.

编织
weaving

支撑
reciprocating

多边形
polygon

宫崎花园露台酒店
Garden Terrace Miyazaki

位置：日本宫崎县

时间：2010—2012

主要用途：酒店

占地面积：10,034.47m²

总建筑面积：4,560m²

结构顾问：福冈耕造（Fukuoka Kouzou）（设计）、牧野结构设计（Makino Structural Design）（监理）

建筑施工：株式会社大林组九州分公司（Obayashigumi Kyushu Branch）

在这个项目中，我希望竹制的建筑外观与以竹子为主体的景观之间形成一种呼应。我们尝试将构成景观与建筑的两种粒子在密度和尺寸上统一，以消解建筑与景观之间的界限。中庭部分做成水面和竹林的组合，水的透明感与竹子的透明感也遥相呼应。

In this design, I wanted to achieve resonances between a bamboo facade and a landscape where bamboo is a main constituent. By unifying the size and density of the particles that make up the landscape and the building, we attempted to minimize the boundary between the landscape and architecture. In the courtyard portion of the building, we juxtaposed a water feature basin with a bamboo grove to create resonances between the different transparencies of water and bamboo.

粒子化
particlizing

包装
wrapping

网格
grid

Shizuku by Chef Naoko餐厅

Shizuku by Chef Naoko

位置：美国俄勒冈州波特兰市

设计时间：2015.10—2016.02

建造时间：2016.03—2016.10

主要用途：餐厅

占地面积：329.45m²

总建筑面积：109.5m²

当地建筑师：罗林·格思里（Lorraine Guthrie）

景观设计：定文内山（Sadafumi Uchiyama）

建筑施工：Kems木工（Kems Woodworks）

在这个项目中，我们将通常竖直悬挂的竹帘旋转90°，然后挂在屋顶上，用半透明的曲面柔和地划分空间。如此，我们在竹帘身上找到了一种新的建筑语言。竹帘的有趣之处在于，虽然竹篾是线形材料，但是由于编织线的介入，可以形成柔软的曲面。竹帘通常的悬挂方法掩盖了它创造曲面的能力——想到这一点，这个新的用法就诞生了。

In this project, we took off-the-shelf sudare (bamboo blinds) and, rotating them 90 degrees relative to their conventional orientation, hung them from the ceiling. This afforded a new means of using curved, semi-transparent surfaces to create gentle divisions. Sudare are interesting because interventions with strings can make their thin bamboo strip—straight, linear members—form gently curving surfaces. This new detail was born when I realized that the conventional hanging method of the blinds ignores their remarkable ability to create curves.

编织
weaving

螺旋
spiral

银山温泉·藤屋
Ginzan Onsen Fujiya

位置：日本山形县尾花泽市
设计时间：2002.04—2005.03
建造时间：2003.04—2006.07
主要用途：日式酒店
占地面积：558m²
总建筑面积：927m²
结构顾问：中田捷夫及合伙人（K.Natata & Associates）
建筑施工：爱和建设株式会社（Aiwa Construction）

在这个项目里，我们用到了一种名为"帘虫笼"的传统编帘工艺——用剖得极细的竹篾编成半透明的屏风，既可以柔和地划分内部空间，还能替代外窗上的窗帘。这种屏风的细腻精致是木格子乃至普通竹百叶窗无法比拟的，这要归功于竹子特有的柔韧性。我们仔细把控竹节位置的对齐和错位，最终获得了一种自然随机的感觉。

In this project, we adapted a detail from sumushiko, traditional louvers made with thin strips of bamboo, to create semi-transparent screens. We used these to gently divide adjacent spaces and as substitutes for curtains in openings in perimeter walls. The delicacy of the screens, which cannot be matched by wood lattices or natural bamboo louvers, was achieved thanks to the flexibility of the bamboo material. By carefully controlling the alignment and misalignment of the bamboo nodes, we created a natural sense of randomness.

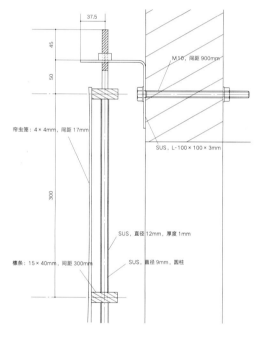

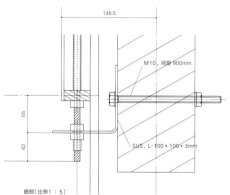

细部（比例1：5）

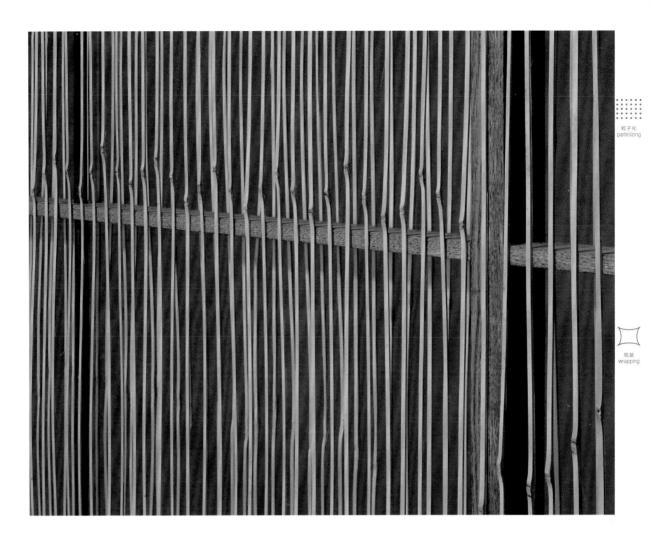

"帘虫笼"是我们在金泽市的民居建筑上发现的一种传统工艺,藤屋里的屏风是金泽市的中田建筑工房制作的。

Sumushiko blinds that consist of a detail found in traditional town houses in Kanazawa were used in the Nakata Construction Workshop in Kanazawa.

竹 / 纤维
Bamboo / Fiber

位置: 东日本
设计时间: 2008.08—2009.04
建造时间: 2009.05—2010.05
主要用途: 私人住宅
占地面积: 459.15m²
总建筑面积: 224.08m²
结构顾问: 江尻建筑结构设计事务所
建筑施工: 松本建筑工程有限公司
（Matsumoto Corporation）

我们把玻璃钢中的玻璃纤维替换为竹纤维，得到了一种色调温暖、具有透光性的板材，并将其应用在屋顶和室内的透光墙面上。室内的梁架采用了竹纤维的抗压构件，这种材料的强度比钢还高。

By substituting the glass fiber of GFRP (glass-fiber reinforced plastic) with bamboo fiber, we created translucent, warmly colored panels and used them in the roof and luminous walls of the interior. We also made beams on the interior using compression members of bamboo fiber that have a higher strength than steel.

内墙是用竹篾编成的网板。竹篾之间的连接不用麻绳，改用细尼龙线和双面胶。

Bamboo lathing made with split bamboo was braided and used as is for the interior walls. The joints were connected with thin nylon thread and double-sided tape, rather than using hemp twine.

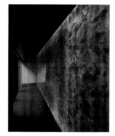

包装
wrapping

网格
grid

以心传心
Nangchang-Nangchang

位置：韩国光州
时间：2013
主要用途：装置
总建筑面积：72m²

着眼于竹子这种材料的柔韧性，我们做了一个会晃动的装置。它能对人的动作做出交互式回应——踩上竹地板，竹地板会将力传递到竹子的两端，越往末端动得越强烈，从而导致整个空间的震动。为了强化这种现象的表现力，我们与艺术家河川吉合作，做了一个把捕捉到的震动转化为声音的音响系统。

就像传统乐器尺八（日本的一种五孔竹笛）一样，这一装置在视觉和听觉方面，为竹子与人创造了一种深层的联系。

包括地板在内的所有构件均用竹子制作。京都的横山竹材店做了实验模型，装置最终由韩国的竹艺大师黄仁真负责制作。

All members were made with bamboo, including the base. A mockup was made at the Yokoyama Bamboo Products in Kyoto, and the final version was made by In-jin Hwang, a master bamboo craftsman in Korea.

Focusing on the flexibility of bamboo, we created
a moving pavilion that responded interactively
to movements. When someone stepped on the
bamboo floor, it transmitted forces to the ends of
the bamboo, amplifying movements at the ends and
causing the entire space to vibrate. We reinforced
this phenomenon by collaborating with artist Yoshi
Horikawa to incorporate a system that picked up
vibrations and translated them into sounds.

The pavilion used bamboo to create deep connections
between people, both visual and aural, as is also true
of shakuhachi (Japanese bamboo flutes).

粒子化
particlizing

编织
weaving

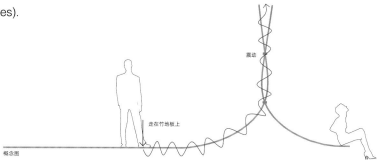

震动

走在竹地板上

概念图

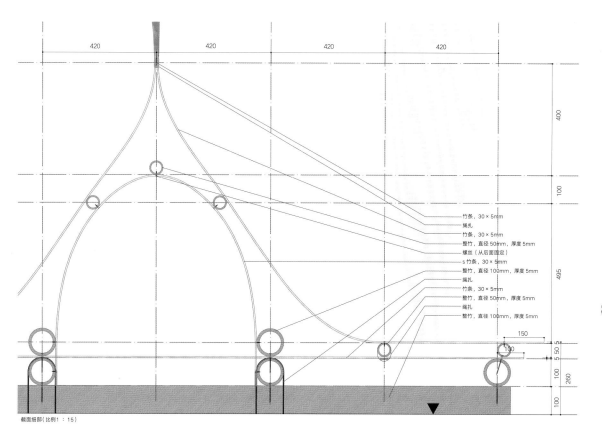

竹条，30×5mm
绳扎
竹条，30×5mm
整竹，直径50mm，厚度5mm
螺丝（从后面固定）
s 竹条，30×5mm
整竹，直径100mm，厚度5mm
绳扎
竹条，30×5mm
整竹，直径50mm，厚度5mm
绳扎
整竹，直径100mm，厚度5mm

截面细部（比例1：15）

螺旋
spiral

感知空间
Sensing Spaces

位置：英国伦敦皇家艺术学院
时间：2014
主要用途：装置

应英国伦敦皇家艺术院的邀请，我们做了这个会散发气味的装置。这个云一般透明的结构体是由直径4mm的竹丝弯曲编成的，用热塑性树脂材料连接。竹丝从地板下的容器内吸收香味（榻榻米的草香和桧木的木香），再将气味散发到整个空间里。我们还尝试改变竹丝的密度，来调节香气的浓淡层次。

At the invitation of the Royal Academy in London, we created this transparent structure resembling a fragrance-emitting cloud. It was woven from bent bamboo strips, 4 mm in diameter, that were joined with thermoplastic resin. The strips absorbed tatami and hinoki fragrances from vessels beneath the floor and suffused it into the space. We attempted to create gradational changes in the fragrance concentration by altering the density of the bamboo strip arrangement.

编织
weaving

支撑
reciprocating

螺旋
spiral

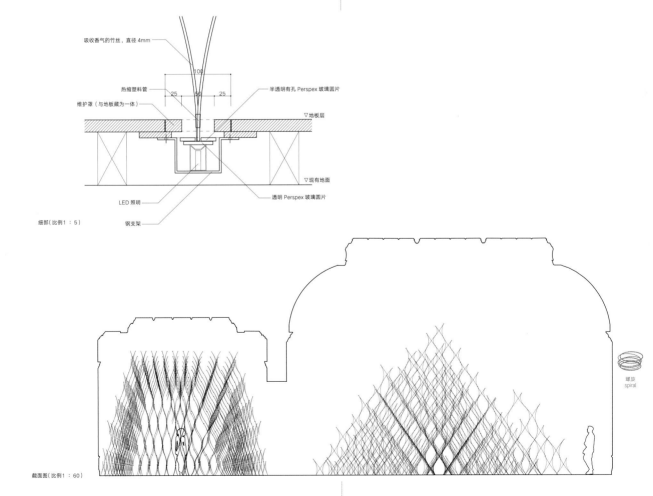

吸收香气的竹丝，直径 4mm

热缩塑料管

维护罩（与地板藏为一体）

半透明有孔 Perspex 玻璃圆片

▽地板层

▽现有地面

LED 照明

透明 Perspex 玻璃圆片

钢支架

细部（比例1∶5）

截面图（比例1∶60）

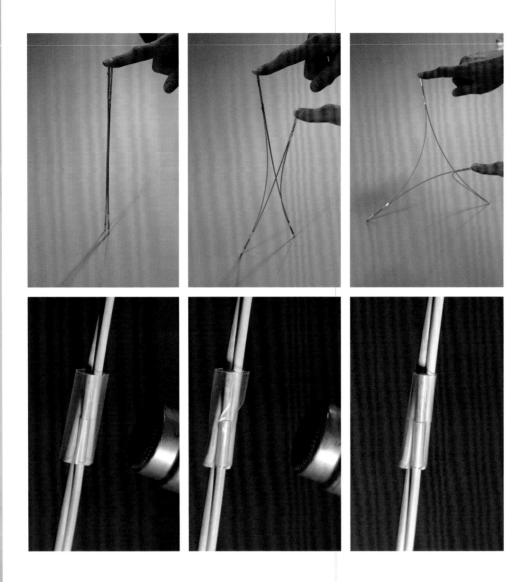

用热气枪使电线的热收缩管收缩，以固定竹丝。长冈造型大学的江尻宪泰先生做结构设计，通过对竹子施加张力以获得足够的强度。

Heat shrink tubing for electrical wiring was shrunk with a heat gun in order to secure the bamboo pieces. The structure was designed by Norihiro Ejiri of the Nagaoka Institute of Design, and adequate strength was obtained by applying tension to the bamboo.

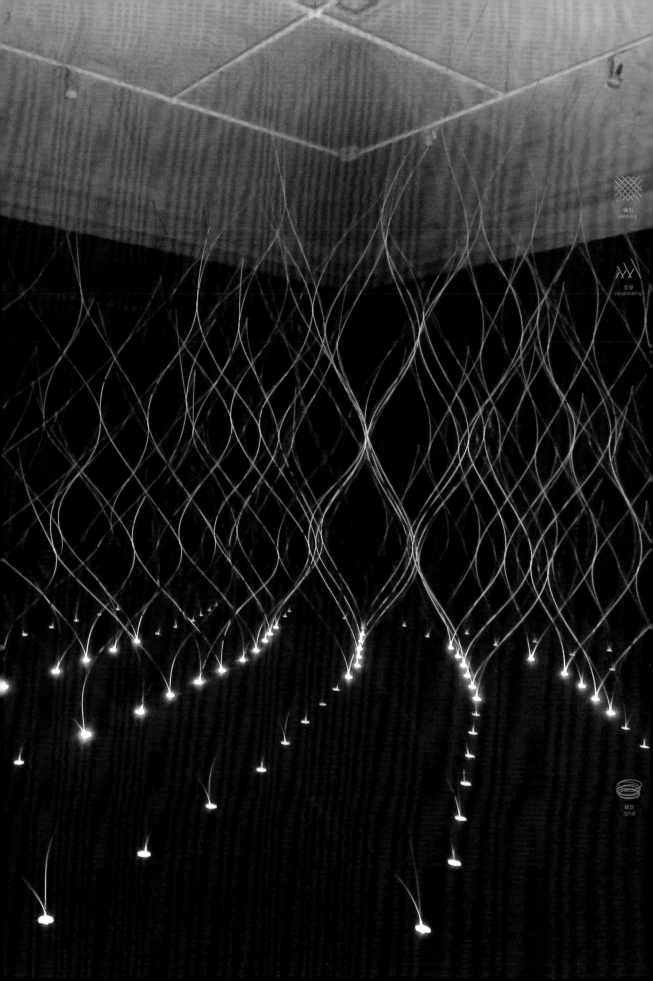

编织
weaving

支撑
reciprocating

螺旋
spiral

Wood

我在横滨的老家是一栋建于20世纪30年代的木结构平房，而且是木结构全都外露的那种做法，非常简朴。我曾经拆掉壁橱上部的天花板，钻到屋架空间里玩耍。

这栋房子不是风雅古典的数寄屋建筑，也不是第二次世界大战后那种用新装饰材料把梁柱遮挡起来的样式，而是典型的战前木结构民居，主要结构用的都是直径100mm左右的木材。对我来说这是件幸运的事，它让我很小的时候就接触到了那个用小径木材编织出来的精巧灵活的建筑世界。

19世纪重要的建筑理论家戈特弗里德·森佩尔认为，建筑是由三个部分的工作组成的：一是关于"地"的工作（可以理解为基础工程），二是关于"火"的工作（可以理解为空调、给排水、电气设备，或可称为设备工程），三是关于"编织"的工作。

森佩尔把基础以外的工作概称为"编织"的工作。当时欧洲主要的建筑方法是垒砌砖石，提出"编织"的概念是需要很大勇气的。一种说法是，因为他父亲开纺织厂，所以他对编织这种工作比较熟悉。但是我更倾向于另一种说法，森佩尔当时参与了世博会的工作，对欧洲以外的传统房屋的建筑方法有所了解，例如亚洲、非洲的传统房屋大多是以树木的枝叶为主要材料"编织"

The house where I grew up in Yokohama was a cheap one-story wooden house built in the 1930s, with minimal framework and walls that left the posts exposed. I used to climb up through a hole in the ceiling of a closet and play in the frame space under the roof.

This house was neither a refined sukiya-zukuri residence nor a postwar-style wooden house with new materials to hide the posts and beams. It was a typical prewar wooden house with a frame of narrow-diameter timber (about 100 mm square). This was lucky for me, because from a young age I was able to come into contact with the delicate and flexible world of architecture woven with narrow-diameter timber.

Gottfried Semper, the most important architecture critic of the 19th century, recognized that architecture is built through three processes. One is related to the earth (it could probably be called foundation work). Another is related to fire (air-conditioning, plumbing, wiring, and other services), and the last is weaving.

Semper used "weaving" as a general term for all work above the foundation. In Semper's time, the main European construction method was brick and stone masonry, so it took courage to use the term "weaving". According to one theory, Semper was acquainted with weaving because of his father's textile factory. But I prefer another theory, which is that he was a man of wide learning who had created designs for the first great international exhibition and studied primitive dwellings outside Europe, and that he used the term "weaving" to convey his understanding of

1

起来的，因此提出了"编织"这个概念。

日本的传统木结构房屋首先是用小径木材编织成框架，然后在框架之间的竹编底板上涂以泥灰，就成了墙。我老家的房子，这种土墙做得很差，泥灰经常掉下来，散落在榻榻米上。刮开土墙，就露出竹篾编成的底板，很容易看出这房子是"编织"出来的。

木材是最容易编织的材料。在木材上做出缺口，缺口与缺口嵌合起来就能固定。日本的传统木结构建筑是避免使用容易生锈的金属固件的。编织的优秀之处是组织的灵活性，这使得衣服能够跟随身体活动。编织木材，我们也追求灵活的连接。木材会膨胀收缩，会翘，会扭曲，灵活的连接才能适应种种变化。

此外，木材很轻却有强度。木质门窗等构件灵活地固定在建筑上，可以自由地适应建筑的变化。

门窗这些构成建筑的粒子灵活放松地连接在一起，就像自由松弛的云。有了木材，我们可以梦想云一般的未来建筑。

their methods of construction. In fact, houses in Asia and Africa are made by weaving, with the branches and leaves of trees as their main materials.

In traditional Japanese wooden houses, the frame was erected first, and then wattle and daub walls were made by applying earthen plaster to takekomai bamboo wattles between the posts. In our house, the earth walls were poorly finished and tended to crumble into dirt underfoot on the tatami. Brushing away the earthen plaster would reveal the wattle made by weaving strips of bamboo. It was easy to see that the house itself was woven.

Wood is the easiest material to weave. It can be fixed in place by cutting notches into each piece and aligning the notches.

Screws and other metal fixtures are avoided in Japanese wooden houses because they tend to rust. The marvelous thing about woven materials is their flexibility, which is what allows clothes to follow the movements of the body. When we weave wood, we try to make the joints flexible in the same way. Wood can expand, shrink, warp, and twist, so the joints must be flexible enough to follow it.

Wood is also light and strong. Units like sliding doors and screens can be made of wood and fixed in place with flexible joints for architecture that freely changes shape.

On the same principle as the sliding doors and screens, architecture could be like a cloud of loosely connected particles. With wood we can dream of futuristic, cloud-like architecture.

2

3

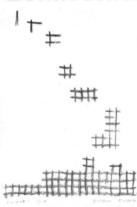

4

5

Sketches by Kengo Kuma
1. Nagaoka City Hall Aore
2. EPFL ArtLab
3. Nakagawa-machi Bato Hiroshige Museum of Art
4. Cidori
5. Asakusa Culture Tourist Information Center
6. Sunny Hills Japan

隈研吾手稿
1. Aore 长冈市政厅
2. 洛桑联邦理工学院艺术实验室
3. 那珂川町马头广重美术馆
4. 千鸟格
5. 浅草文化观光中心
6. 微热山丘・日本店

6

那珂川町马头广重美术馆
Nakagawa-machi Bato
Hiroshige Museum of Art

位置：日本栃木县

时间：1998—2000

主要用途：日本艺术博物馆

占地面积：5,586.84m²

总建筑面积：1,962.43m²

结构顾问：青木结构顾问事务所

（Aoki Structural Engineers）

建筑施工：株式会社大林组（Obayashi Corporation）

含氟聚合物涂层钢，厚度 0.6mm

屋顶：波形夹丝玻璃

玻璃屋顶支撑：M10

连接螺栓：12 套管，17.3 镀锌，SOP

镀锌表面 SOP 封装金属附件：St L-50 × 70mm，厚度 6mm

隔栅墙

店前花园

含氟聚合物涂层钢板，直立缝，宽度 0.6mm

截面图（比例 1：50）

广重美术馆是一座整体覆盖着木百叶窗的建筑，这
是歌川广重（1797—1858）浮世绘作品中的雨给
我带来的灵感。广重"粒子化"的艺术手法对弗兰
克·劳埃德·赖特设计的各种屏风，以及凡·高等
印象派和后印象派画家的点彩画法也产生过影响。

百叶窗的基本构件是经过防火处理的30×60mm
规格的杉木条，按120mm的间隔排列。地面铺贴
的石材和门窗隔扇的分割也都以120的倍数为单位。
室内天花板和屋顶双层铺贴与百叶窗相同的杉木条，
营造出一种树林中漏下阳光的光影效果。木材的防
火处理技术是由宇都宫大学的安藤实先生研发的。

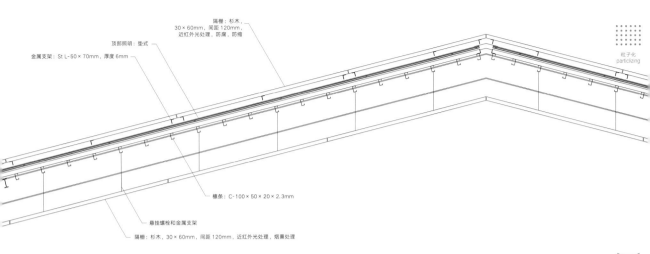

隔栅：杉木，
30×60mm，间距 120mm，
近红外光处理，防腐，防缩

顶部照明：垫式

金属支架：St L-50×70mm，厚度 6mm

檩条：C-100×50×20×2.3mm

悬挂螺栓和金属支架

隔栅：杉木，30×60mm，间距 120mm，近红外光处理，烟熏处理

粒子化
particlizing

包装
wrapping

The depiction of rain in ukiyo-e prints by Utagawa Hiroshige (1797—1858) influenced my decision to cover this entire building in wood louvers. Hiroshige's "particlizing" techniques also influenced Frank Lloyd Wright in his designs of various screens and Van Gogh and the Impressionists in their use of pointillism.

We made the louvers from pieces of Japanese cedar, 30 mm by 60 mm in section, that have undergone fireproofing treatment. The louvers are spaced at intervals of 120 mm, a measurement that serves as the basic unit of the design. The location of the stone paver joints and the layout of the shoji are based on multiples of the 120 mm unit. Fireproofed cedar louvers with the same cross section (30 x 60mm) were attached to the roof and ceiling in double layers to obtain the effect of sunlight filtered by trees. Minoru Ando of Utsunomiya University developed the fireproofing technology.

网格
grid

千鸟格
Cidori

位置：意大利米兰

时间：2007

主要用途：装置

总建筑面积：15m²

结构顾问：佐藤淳构造设计事务所
（Jun Sato Structural Engineers）

建筑施工：KAYA工作室（studio KAYA）

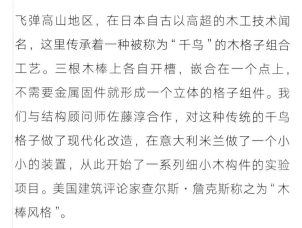

飞弹高山地区，在日本自古以高超的木工技术闻名，这里传承着一种被称为"千鸟"的木格子组合工艺。三根木棒上各自开槽，嵌合在一个点上，不需要金属固件就形成一个立体的格子组件。我们与结构顾问师佐藤淳合作，对这种传统的千鸟格子做了现代化改造，在意大利米兰做了一个小小的装置，从此开始了一系列细小木构件的实验项目。美国建筑评论家查尔斯·詹克斯称之为"木棒风格"。

In Hida-Takayama, an area known for technically sophisticated carpentry, a wood joinery system known as cidori has been passed down since ancient times. Under the system, notches are made at points along straight wooden members in a way that enables three such members to be joined at each of the points. The members are assembled into three-dimensional lattices that do not require hardware. In collaboration with structural engineer Jun Sato, we modernized the cidori system and used it to create a small pavilion in Milan, Italy. This was the first in a series of projects using slender wood members-projects that architectural historian Charles Jencks has termed" stick style."

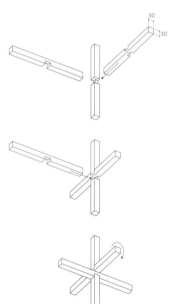

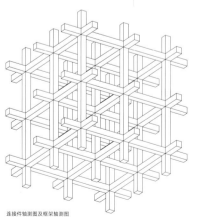

连接件轴测图及框架轴测图

在 30 × 30mm 的桧木条上开槽，相互拧紧固定。

Notches were cut in 30 × 30 mm Japanese Cypress pieces, and they were secured to each other by twisting them.

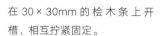

GC 齿科博物馆研究中心
GC Prostho Museum
Research Center

这是我们将千鸟格系统应用到永久建筑上的首次尝试。在这里，木棒不仅作为结构材料支撑着建筑，也是装置和展示架：4个单元格厚（2m）的"墙壁"可以直接用来陈设展品。把一个单元复制、扩张到世界级的规模，我们把这个想法叫作"细胞的世界观"。20世纪的生物学和现代主义建筑是以器官为单位来理解生物和建筑的，现在，我们以细胞为单位，不以器官为媒介，营造一个属于细胞的世界。

位置：日本爱知县

时间：2008—2010

主要用途：博物馆、研究中心

占地面积：421.55m²

总建筑面积：626.5m²

合作建筑：松井建设株式会社设计部

（Design Department of Matsui Construction）

结构顾问：佐藤淳构造设计事务所

建筑施工：松井建设株式会社

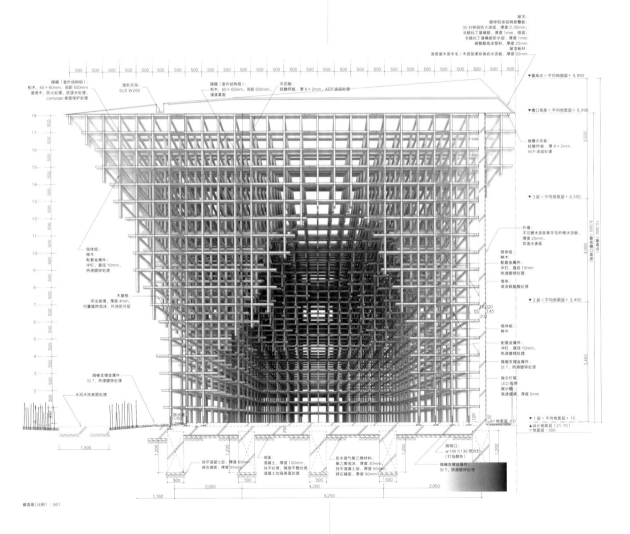

截面图 [比例1：80]

This was the first project in which we developed the Cidori's grid frame system for use in a permanent building. Here, the sticks are not only structural members supporting the building but also furniture and display cases. Walls with a thickness of four grids (2 meters) were deep enough to be used as display cases, without modification. We refer to the idea of replicating a single unit until it reaches the scale of the world as a "cellular world view." Whereas the 20th century biologists and modernist architects understood living things and architecture based on organ units, we strive to understand them based on cellular units that can achieve a world themselves, unmediated by organs.

粒子化
particlizing

编织
weaving

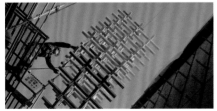

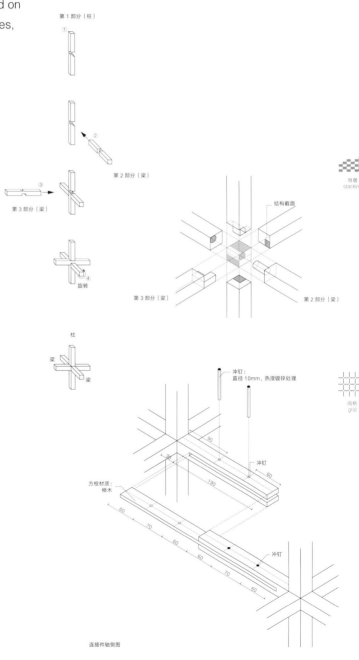

第 1 部分（柱）

第 2 部分（梁）

第 3 部分（梁）

第 3 部分（梁）

第 2 部分（梁）

结构截面

柱
梁
梁

冲钉：
直径 10mm，热浸镀锌处理

冲钉

冲钉

方栓材质：
榉木

冲钉

堆叠
stacking

网格
grid

连接件轴侧图

星巴克咖啡·太宰府
天满宫表参道店
Starbucks Coffee at
Dazaifutenmangu Omotesando

位置：日本福冈县
设计时间：2011.01—2011.08
建造时间：2011.08—2011.11
主要用途：咖啡店
占地面积：436.71m²
总建筑面积：210.03m²
结构顾问：佐藤淳构造设计事务所
建筑施工：松本建筑工程有限公司

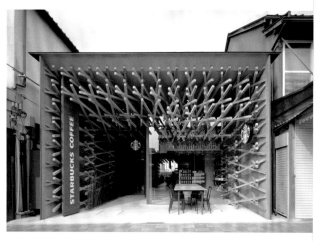

在这个项目中，我们将在米兰及"GC齿科博物馆研究中心"做的直交格子转换成了斜交的立体格子。直交的格子是静态的，而我们要做的临街店铺开口小、进深细长，斜交格子的设计能带来一种不断向内部深入的流动感。直交格子是三根木棒组成一个单元，斜交格子则是四根木棒交错组成一个单元，嵌合更加复杂。木棒与齿科博物馆研究中心粗细相同，都是60×60mm规格的，但在连接上用到了不锈钢的销子。

In this project we took the grid frame system of Cidori and GC Prostho Museum Researth Center and converted it into an oblique three-dimensional lattice. Orthogonal grids have a static appearance. By making this grid oblique, we gave it a sense of fluidity to help coax people into a deep space with a narrow frontage, typical of shops along Omotesando's promenade. Whereas orthogonal grids have three wooden members crossing at intersections, this oblique grid has four, which necessitated notches of greater complexity. The sticks are 60 mm square in section, just like those of Prostho Museum. The joints, however, rely on stainless steel dowel pins.

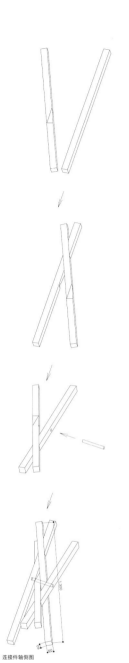

连接件轴侧图

粒子化
particlizing

编织
weaving

包装
wrapping

多边形
polygon

结构全部都由 60×60mm 规格的桧木条组合而成，材料长度总计有 4,390m。

Pieces of Japanese Cypress that all have a cross section of 60×60 mm were assembled. The total length of the members is 4,390 meters.

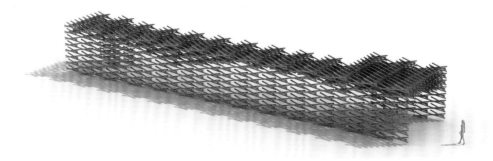

微热山丘·日本店
Sunny Hills Japan

位置：日本东京
设计时间：2012.01—2012.10
建造时间：2012.11—2013.12
主要用途：零售店
占地面积：175.69m²
总建筑面积：293m²
结构顾问：佐藤淳构造设计事务所
建筑施工：Satohide公司（Satohide Corporation）

在这个项目中，我们把"星巴克咖啡·太宰府天满宫表参道店"的斜交立体格子转换了一下方向，做成一个能够支撑起三层建筑的分层结构体。常见的框架结构建筑就像是内部由骨骼支撑的脊椎动物，而这种外部的网状结构就像昆虫的坚硬外皮，可以称作"昆虫型结构"。

支撑一座三层的永久建筑，木组件需要具备比星巴克项目更高的强度。我们借鉴了日本传统的组木工艺"地狱组"的做法，成功做出了一个三层建筑的木组结构。

"地狱组"是一种给门窗、家具等加固的三层格子的组合方式，两层的格子会产生滑动移位，再加上一层就能阻止滑动，获得结构上的刚性。

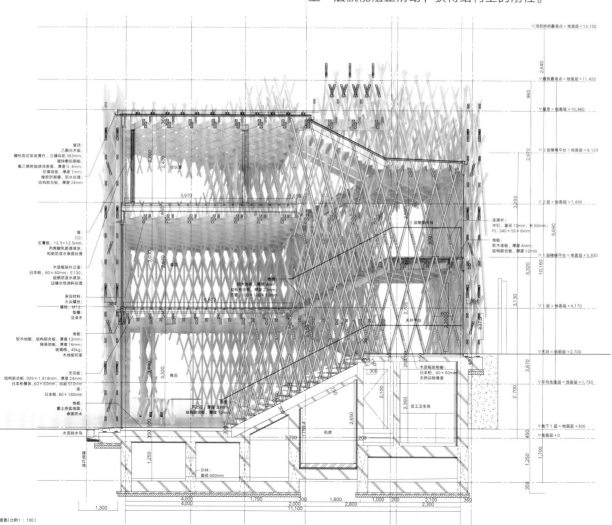

截面图（比例1：100）

By altering the direction of the oblique three-dimensional lattice of Starbucks Coffee at Dazaifutenmangu Omotesando, we transformed the lattice into a layered structural system that supports a three-story building. Whereas ordinary, rigid frame structures have internal skeletons like vertebrates, the mesh structure of Sunny Hills Japan has a rigid, mesh-like outer skin, and could rightly be characterized as an" insect structure."

Supporting a permanent, three-story building, required stronger joints than the ones we used in Starbucks. Drawing upon a traditional Japanese structural system called jigoku-gumi, or "hell joinery," we settled on a three-layer timber structure.

Jigoku-gumi, which was originally used to strengthen architectural fittings, furniture, and so on, is a three-layered lattice system. Whereas two-layered lattices are susceptible to slippage, the addition of a third layer prevents slippage and ensures frame rigidity.

粒子化
particlizing

编织
weaving

包装
wrapping

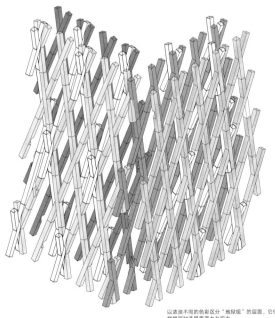

多边形
polygon

以浓淡不同的色彩区分"地狱组"的层面,它们
能够同时承受垂直力与扭力。
Parallel layers of Jigokugumi (same hue) work in
pairs; they can receive both vertical loads and
tortional forces.

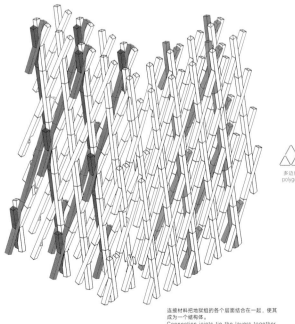

连接材料把地狱组的各个层面结合在一起,使其
成为一个结构体。
Connection joints tie the layers together,
allowing them to work as one structure.

做立体的"地狱组"时，首先用 3D 软件在电脑中模拟，然后用实物大小的泡沫材料做模型实验，最后再用桧木材料做最终的结构实体。

Modeling for the 3D Jigoku-gumi structure was first performed with 3D software, a mock-up was built using actual size styrofoam square members, the assembly methods were evaluated, and then Japanese Cypress was used to make the final mock-up.

两件组   四件组

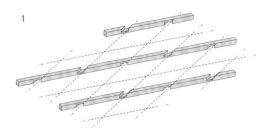

1

第一步，平行摆放 X 方向的木条，木条数量为底层的一半（木条上下间隔着开槽，开槽深度为木条总厚度的 2/3）。
The first half of the bottom-layer of battens is placed on the grid (X direction). The depth of the notches is 2/3rds of their total thickness.

2

第二步，将 Y 方向的木条对准上下凹槽嵌入，木条数量为上层的一半（木条的开槽深度同前）。
The first half of the top-layer of battens is placed on the grid (Y direction). The depth of the notches is 2/3rds of their total thickness, and they interlock with those of the bottom layer.

3

第三步，X 方向的另一半木条也对准上下凹槽嵌入（木条的开槽深度同前），然后将第一步与第三步的 X 方向的木条都错开 1/3，得到空隙可使 Y 方向剩余的一半木条插入。
The second half of the top layer is placed on the grid (X direction). The depth of the notches is 2/3rds of their total thickness, and they interlock with the other half of the top layer. This system leaves 1/3rd of thickness free, in order to manouver the battens up and down to insert the last layer.

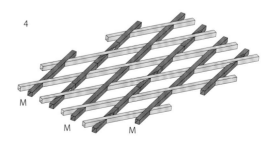

4

M   M   M

第四步，将 Y 方向剩余的一半木条插入（木条的开槽深度为 1/3），然后将 X 方向的木条复位。
The layers are shifted up and down by 1/3rd of their thickness, allowing the last "Master layer" to be inserted (Y direction), and lock all battens together.

晃动
Yure

位置：法国巴黎

设计时间：2015.08—2015.10

建造时间：2015.10

主要用途：装置

占地面积：35m²

高：11.6m

结构顾问：江尻建筑结构设计事务所

建筑施工：雷米·贝格尼斯（Remy Begnis）、
巴贝特·加利安（Babeth Galian）

在星巴克咖啡、微热山丘之后，我们又做了这个斜格子的装置。之前一系列的项目用的都是60×60mm正方形截面的木材，这次我们改用90×180mm的长方形截面的木材，使材料带有方向性，从不同的方向看过去会有不同的透明感。这个装置不仅仅因为线形材料斜向组合获得了结构上的强度，还造成了一种螺旋上升的动态，给巴黎宁静的杜伊勒里花园带来了"晃动"。

We created this temporary monument using linear members arranged in an oblique three-dimensional lattice, as we did in Starbucks Coffee at Dazaifutenmangu Omotesando and Sunny Hills Japan. Here, however, we used sticks that are rectangular in section (90 mm by 180 mm) instead of the 60 mm square section sticks of the earlier projects. Rectangular section sticks have directionality. This enabled us to modulate the transparency of the structure based on the direction from which it was viewed. The linear members joined obliquely in a braced structure provided strength. In addition, they created yure—the Japanese word "rocking" or "undulating" —in the static Parisian plaza of Tuileries Garden by drawing people up into a giant spiral that rose toward the heavens.

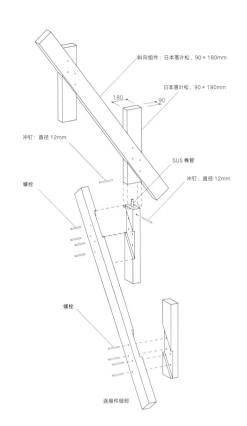

斜向组件：日本落叶松，90×180mm

日本落叶松，90×180mm

180

90

冲钉：直径12mm

SUS 椎管

冲钉：直径12mm

螺栓

螺栓

连接件细部

粒子化
particlizing

编织
weaving

堆叠
stacking

多边形
polygon

螺旋
spiral

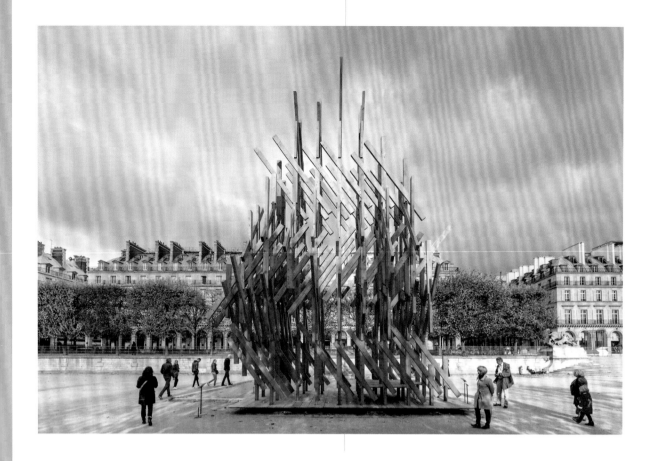

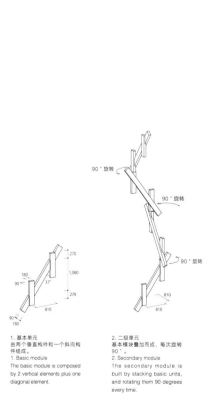

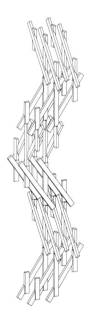

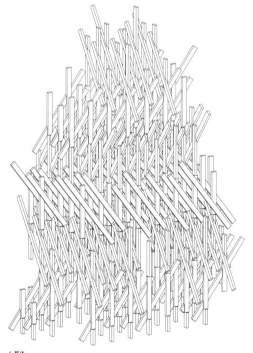

1. 基本单元
由两个垂直构件和一个斜向构件组成。
1. Basic module
The basic module is composed by 2 vertical elements plus one diagonal element.

2. 二级单元
基本模块叠加而成,每次旋转90°。
2. Secondary module
The secondary module is built by stacking basic units, and rotating them 90 degrees every time.

3. 网格系统
二级模块沿着X、Y、Z方向复制,成为可以无限反复的网格系统。
3. Grid system
The secondary modules can be repeated along the XYZ axes, and aggregate into a three dimentional lattice.

4. 整体
基本单元与二级单元通过加减成为一个整体,垂直材料与斜向材料的长度有时延长有时缩短,形成不规则的轮廓。
4. Pavilion
By adding and removing elements from the three dimentional lattice, an ambiguous volume with irregular outline is created.

粒子化
particlizing

编织
weaving

堆叠
stacking

多边形
polygon

螺旋
spiral

梼原木桥博物馆
Yusuhara Wooden Bridge Museum

位置：日本高知县

时间：2009—2010

主要用途：画廊

占地面积：14,736.47m²

总建筑面积：445.79m²

结构顾问：中田结构设计事务所

建筑施工：Shimanto Sogo建设（Shimanto Sogo Construction）

在这个项目中，我们试图用一系列的小构件来实现一个大悬臂结构。为此，我们对亚洲传统木构建筑中支撑大屋檐的斗拱结构进行了研究。山梨县的山里有一座名叫"猿桥"的木桥，就是用斗拱结构解决了大跨度的问题。在梼原木桥博物馆这个项目里，我们将桥墩改为居中配重的平衡结构，把跨度一分为二。

欧洲的大型木结构建造物主要依赖大截面的集成材，而梼原町的工厂无法生产截面尺寸300mm以上的集成材，因此我们反过来考虑运用小截面材料的可能性，这也是在人性化的层面上对传统木建筑的继承。

In this project, we sought to realize a large cantilever supported by a series of small members. To this end, we researched the tokyo system—a system developed to support the large wooden eaves that typify traditional Asian buildings. In our design, we referred to the structure of Saruhashi (literally, "Monkey Bridge"), a long-span wooden bridge in Yamanashi Prefecture that utilizes the tokyo system, but changed it by introducing a single, centrally located pier. The pier cuts the span in two and assumes a starring role in the design. The inclusion of the pier is based on the structural principle behind yajirobe, traditional Japanese toys that balance on a central point through counterweighting.

Laminated members with a large section play a leading role in large wooden structures in Europe, but laminated members with a cross section of 300 mm or more could not be made at the facility in Yusuhara-cho. This fact which may appear to be a disadvantage was utilized to pursue the potential of small-section lamination in attempt to carry on the tradition of human scale in wooden structures in Japan.

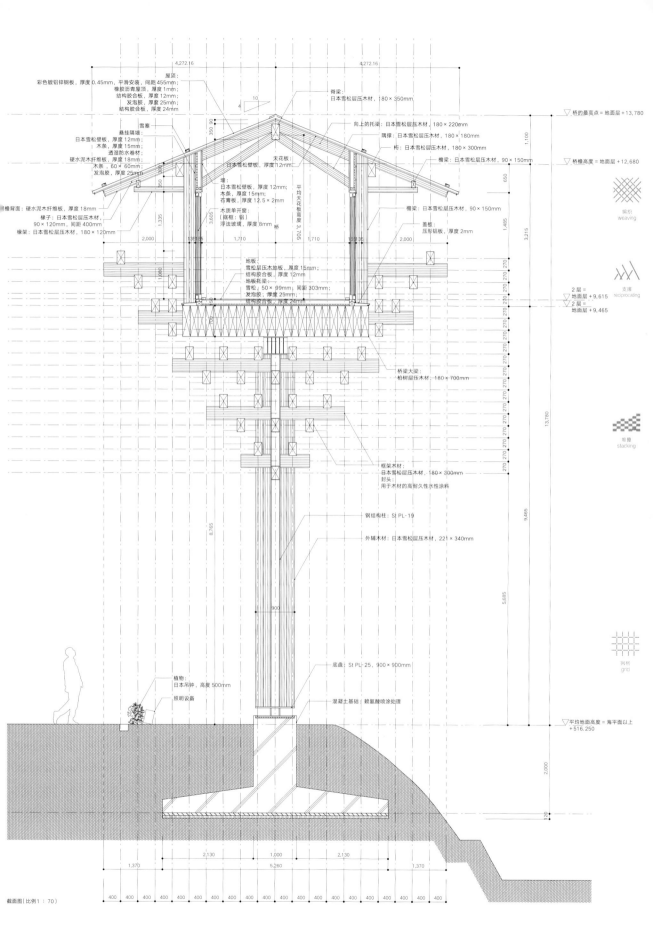

彩色镀铝锌钢板，厚度 0.45mm，平滑安装，间距 455mm；
橡胶沥青屋顶，厚度 1mm；
结构胶合板，厚度 12mm；
发泡胶，厚度 25mm；
结构胶合板，厚度 24mm

屋顶

脊梁：
日本雪松层压木材，180×350mm

向上的托架：日本雪松层压木材，180×220mm

隅撑：日本雪松层压木材，180×180mm

桁：日本雪松层压木材，180×300mm

檐梁：日本雪松层压木材，90×150mm

悬挂隔墙：
日本雪松壁板，厚度 12mm；
木条，厚度 15mm；
透湿防水卷材；
硬水泥木纤维板，厚度 18mm；
木条，60×60mm；
发泡胶，厚度 25mm

天花板：
日本雪松壁板，厚度 12mm

墙：
日本雪松壁板，厚度 12mm；
木条，厚度 15mm；
苫膏板，厚度 12.5×2mm；

木质单开窗：
（底框：铝）
浮法玻璃，厚度 8mm

檐梁：日本雪松层压木材，90×150mm

盖板：
压形铝板，厚度 2mm

檐背面：硬水泥木纤维板，厚度 18mm
椽子：日本雪松层压木材，
90×120mm，间距 400mm
椽架：日本雪松层压木材，180×120mm

平均天花板高度 3,705

桥

地板：
雪松层压木地板，厚度 15mm；
结构胶合板，厚度 12mm
地板托梁：
雪松，50×99mm，间距 303mm
发泡胶，厚度 25mm
结构胶合板，厚度 24mm

桥梁大梁：
柏树层压木材，180×700mm

框架木材：
日本雪松层压木材，180×300mm；
封头：
用于木材的高耐久性水性涂料

钢结构柱：St PL-19

外辅木材：日本雪松层压木材，221×340mm

植物：
日本吊钟，高度 500mm

照明设备

底盘：St PL-25，900×900mm

混凝土基础：赖氨醇喷涂处理

桥的最高点 = 地面层 +13,780

桥檐高度 = 地面层 +12,680

编织
weaving

支撑
reciprocating

2 层 =
地面层 +9,615
2 层 =
地面层 +9,465

堆叠
stacking

钢树
grid

平均地面高度 = 海平面以上
+516.250

截面图（比例 1：70）

編織
weaving

支撐
reciprocating

堆疊
stacking

網格
grid

九州艺文馆（2号别馆）
Kyushu Geibunkan Museum (Annex 2)

位置：日本福冈县
设计时间：2008.09—2011.03
建造时间：2011.03—2012.10
主要用途：陶瓷艺术工作室
占地面积：12,914.74m²
总建筑面积：165.51m²
合作建筑：株式会社日本设计（Nihon Sekkei）
结构顾问：江尻建筑结构设计事务所
建筑施工：田村建设（Koga kensetsu）

风车结构示意。
Example of pinwheel structure.

将薄板互相嵌合，便做成了立体格子状的结构体。为了实现灵活、通透的空间，我们利用六角形的几何特性做出高度刚性的屋顶，只在必要的地方竖一些细木柱做支撑。巴克敏斯特·富勒（1895—1983）曾把那种像风车一样从细柱上伸出梁的结构称为"风车结构"，还用一把叉子来比喻说明。这里，我们用削掉两角的集成板材——形状有点像内裤——替代顶端尖利的叉子，从而实现了富勒提倡的具有浮游感的结构体。

Fitting thin boards together, we applied a three-dimensional lattice structure method to the design of this permanent building. In order to achieve transparent, flexible spaces, we created a highly rigid roof based on hexagonal geometry and supported it with slender columns arranged in an ad hoc manner. Architect and inventor Buckminster Fuller (1895—1983) referred to this kind of structure, with the beams extending from slender columns in pinwheel formations, as a "pinwheel structure" and depicted it with forks. In our design, in place of forks with sharp prongs, we used laminated wood panels with the ends chopped off—panels shaped something like underpants. In so doing, we gave the structure the light, floating quality that Fuller advocated.

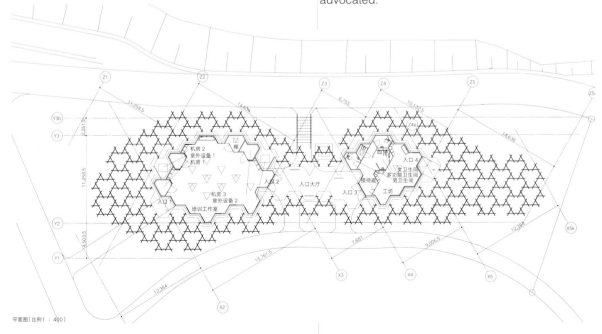

平面图（比例1：400）

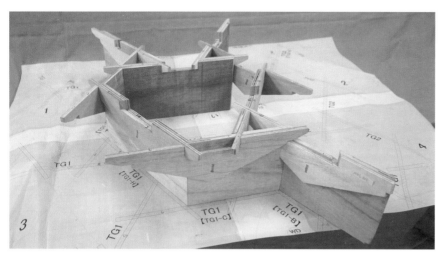

利用六角形的几何特性，实现可以向任何方向自由扩展的建筑。

Hexagonal geometry was used to create a pavilion that can be freely expanded in any direction.

编织
weaving

支撑
reciprocating

多边形
polygon

浅草文化观光中心
Asakusa Culture Tourist Information Center

位置：日本东京
设计时间：2009.01—2010.01
建造时间：2010.08—2012.02
主要用途：旅游信息中心、展厅
占地面积：326m²
总建筑面积：2,142m²
结构顾问：牧野结构设计
建筑施工：藤田／大有设计联合（Fujita/Daiyu JV）

木结构平房曾经是构成东京这个城市的"基本粒子"，于是我设想做一个由8层平房堆叠起来的建筑。20世纪的建筑体系，是相同平面、剖面楼层的单纯堆叠，我对此做了一种反向思考。堆叠的7个"家"的外立面都安装着300mm宽的木百叶窗，各层可以单独调整百叶窗的间距和角度，这样，各层的"家"看起来就像是独立的存在。百叶窗也起到阻隔阳光的作用。

I conceived this mid-rise building as an eight-story stack of single-story wood buildings, once the "fundamental particles" that constituted Tokyo. It is a variation on conventional, the 20th century construction systems, where floors with identical plans and sections were stacked one on top of another. Wood louvers, 300 mm deep, are affixed to each of the seven "house" layers. By adjusting the pitch and angle of the louvers on each layer, we made each "house" seem like an independent entity. The louvers also function as devices that cut sunlight.

叶隙间洒落的阳光 /
拉科斯特城堡
KOMOREBI / Château La Coste

位置：法国普罗旺斯艾克斯拉科斯特城堡

设计时间：2014.08—2015.08

建造时间：2016.06—2017.12

主要用途：装置

结构顾问：江尻建筑结构设计事务所

建筑施工：比约克·穆勒（Bjørk Møller）

在法国南部一处面向圣维克多山的斜坡上，我们用20mm厚的耐久性重蚁木板做了一个屋顶般的装置。木板一点点伸出来，形成一个树样的有机体。需要强度的地方，在木板之间插入4mm的不锈钢板。这种连接方法，加上埋在斜坡里的地基，使得浮游般轻盈的悬臂结构得以实现。某种意义上说，这是一个由薄木板与不锈钢组成的结构体。就像塞尚在绘画中对圣维克多山的抽象化表现，我们做的是树木的抽象化表现。

On a slope facing Montagne Sainte-Victoire in Southern France, we arranged 20-mm-thick pieces of highly durable ipe wood in a way that projects, little by little, and forms an organic, tree-like roof. In areas requiring strength, we used a detail where 4 mm stainless steel plates are sandwiched between wood members. These joints, in combination with a foundation buried in the slope, make the light, cantilevered structure appear to float. Since the structure combines thin wood pieces and stainless steel, it is, in a sense, a mixed structure. We aimed to abstract a tree in the design, just as Cézanne abstracted Montagne Sainte-Victoire in his paintings.

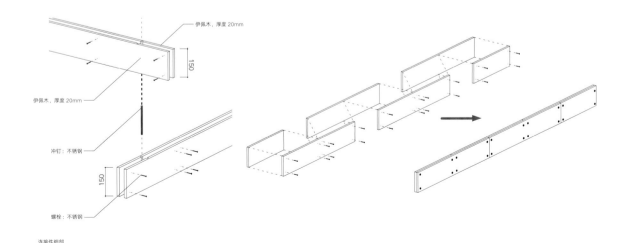

连接件细部

木材与不锈钢的连接方法，是我们与丹麦的木工比约克·穆勒先生一起经过反复实验决定的。

A joint technique using wood and stainless steel sheets was created through a process of trial and error with the Danish carpenter Bjørk Møller.

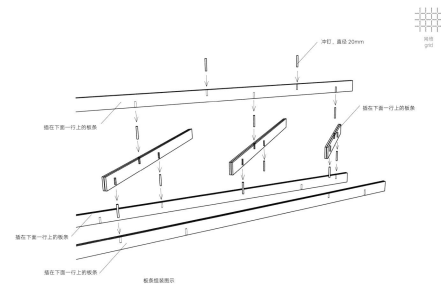

冲钉，直径20mm

插在下面一行上的板条

插在下面一行上的板条

插在下面一行上的板条

插在下面一行上的板条

板条组装图示

网格
grid

Aore 长冈市政厅
Nagaoka City Hall Aore

位置：日本新潟县
时间：2008.02—2012.02
主要用途：市政厅、舞台
占地面积：14,938.81m²
总建筑面积：40,000m²
结构顾问：江尻建筑结构设计事务所
建筑施工：大成建设（Taisei）、福田建设（Fukuda）、
中越建设（Chuetsu）、池田联合（keda JV）

为了赋予市政府的中庭空间宜人的氛围和质感，我们用杉木条做成平板，按照一定角度排列成波浪状，装饰在中庭的立面和天顶上，给雪国带来一丝温暖柔和。

我们觉得，不能仅仅把木材看作一种结构，还要把它当作一种有厚度的真实物体来对待。做这样的设计正是出于这种考虑。为了消除平板背后框架的痕迹，我们将面板直接安装在金属网上，这个方法非常有挑战性。

To give this city hall courtyard space a human scale and texture, we made panels from planks of Japanese cedar and lined them up at angles, in zigzag formation. In so doing, we created a ceilings and interior elevation with warm, soft textures suited to the snowy locale.

We arrived at this detail by treating wood not merely as a texture, but as a real object, with thickness. To eliminate the appearance of frames behind the wood planks, we challenged ourselves to affix the planks directly to expanded metal, which we achieved using a complicated detail.

我们规定木材只用现场周边方圆15千米内的越后杉。以随机的尺寸切割木材，保留木材上的节瘤，从而突出木料的质感，即使从远处也能看出来。

A rule was made that only Echigo cedar sourced within 15 km of the site could be used, and pieces with random dimensions were laid out that have many knots to give the feeling of the texture of wood, even when viewed from quite a distance.

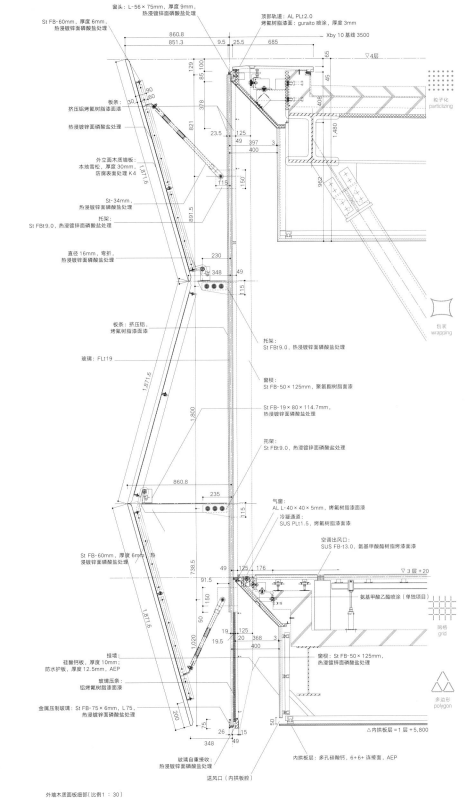

窗头：L-56×75mm，厚度9mm，
热浸镀锌面酸盐处理

St FB-60mm，厚度6mm，
热浸镀锌面磷酸盐处理

顶部轨道：AL PLt2.0
烤氟树脂面漆：guraito 喷涂，厚度3mm

Xby 10 基线 3500

860.8
851.3   9.5 25.5   685   65

▽4层

129   100
85   85

粒子化
particlizing

板条：
挤压铝烤氟树脂面面漆

热浸镀锌面磷酸盐处理

821   378

23.5   125
49   397   3
400

962   1,480

外立面木质墙板：
本地雪松，厚度30mm，
防腐表面处理K4

St-34mm，
热浸镀锌面磷酸盐处理

托架：
St FBt9.0，热浸镀锌面磷酸盐处理

150   115

直径16mm，弯折，
热浸镀锌面磷酸盐处理

230   49
348   115

板条：挤压铝，
烤氟树脂面漆

玻璃：FLt19

1,871.6

1,800

托架：
St FBt9.0，热浸镀锌面磷酸盐处理

窗梃：
St FB-50×125mm，聚氨酯树脂面漆

St FB-19×80×114.7mm，
热浸镀锌面磷酸盐处理

托架：
St FBt9.0，热浸镀锌面磷酸盐处理

860.8
235   115

气窗：
AL L-40×40×5mm，烤氟树脂面漆
冷凝通道：
SUS PLt1.5，烤氟树脂面漆

空调出风口：
SUS FB-t3.0，氨基甲酸酯树脂烤漆面漆

St FB-60mm，厚度6mm，热
浸镀锌面磷酸盐处理

738.5

49   125   176

▽3层 +20

氨基甲酸乙酯喷涂（单独项目）

91.5   150

网格
grid

50   1,020   19   125
19.5   20   368   3
400

挂墙：
硅酸钙板，厚度10mm；
防水护板，厚度12.5mm，AEP

玻璃压条：
铝烤氟树脂面漆

金属压制玻璃：St FB-75×6mm，L75，
热浸镀锌面磷酸盐处理

200

75
26   15
348   49

玻璃自重接收：
热浸镀锌面磷酸盐处理

送风口（内拱板腔）

外墙木质面板细部 [比例1：30]

多边形
polygon

50

△内拱板层 =1层 +5,800

内拱板层：多孔硅酸钙，6+6+ 连接面，AEP

包装
wrapping

我们用错落的杉木板做了模型，也向公众展示。
We built a mock-up of cedar panels in a zigzagged (cidori)
formation and showed it to the public.

东京大学研究生院·大和普适计算研究大楼
Daiwa Ubiquitous Computing Research Building

位置：日本东京

设计时间：2010.09—2012.10

建造时间：2012.10—2014.04

主要用途：大学设施、美术馆、会议室、咖啡厅

占地面积：304,880.06m²

总建筑面积：2,709.53m²

结构顾问：校园规划办公室（隈研吾）及

设备科[Campus Planning Office（Kengo Kuma）& Facilities Department]、大和房建（Daiwa House Industry）

建筑施工：大和房建

完成"Aore长冈市政厅"之后，我们再次尝试用木材做一个有机的、以人为本的立面。在"Aore长冈市政厅"项目里，我们将木板排列成波浪状，这里，我们把它们像鱼鳞一样叠起来，并做成从剖面上看越接近地面越后退的形态。调整鳞片的叠层、错位，可以给广阔的外立面营造出一种壮观的动态效果。

In this project, we attempted to make an organic, human-scaled facade from wood, using plank panels. It is an extension of the facades we designed for Nagaoka City Hall Aore. In Nagaoka City Hall Aore, we set the panels at angles in zigzag formation; here, we layered them like scales and adopted a sectional shape that steps them back as they approach the ground. By adjusting the manner in which the "scales" are layered and shifted, we produced dynamic effects in the building's long facade.

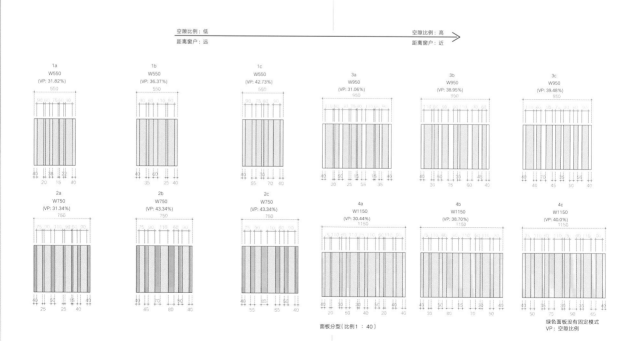

面板分型（比例1∶40）

绿色面板没有固定模式
VP：空隙比例

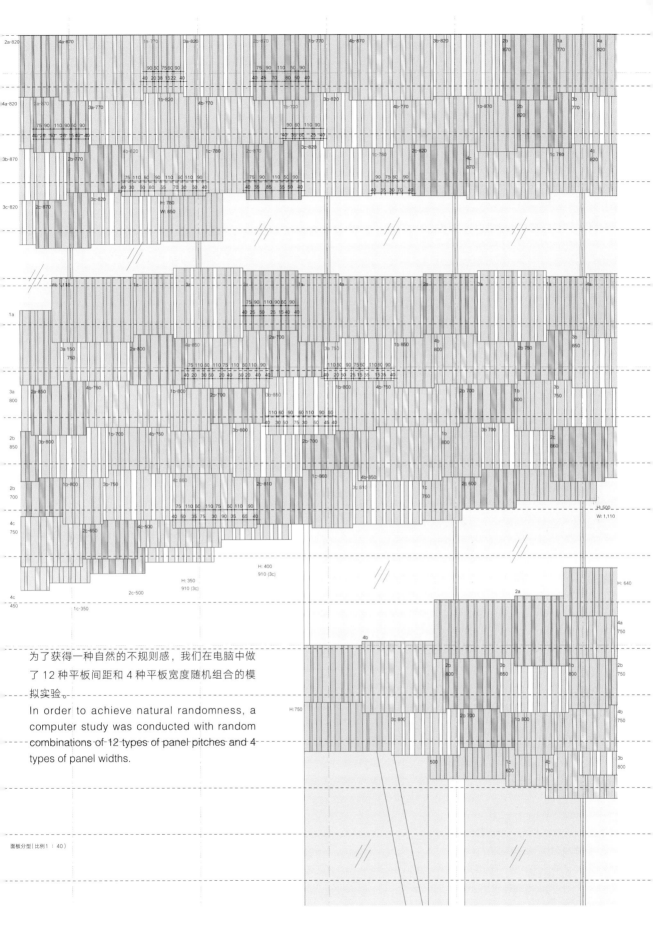

为了获得一种自然的不规则感，我们在电脑中做了 12 种平板间距和 4 种平板宽度随机组合的模拟实验。

In order to achieve natural randomness, a computer study was conducted with random combinations of 12 types of panel pitches and 4 types of panel widths.

面板分型（比例 1：40）

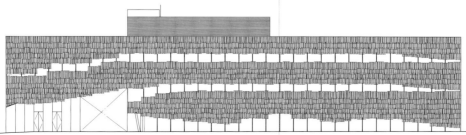

立面图 (比例 1 : 500)

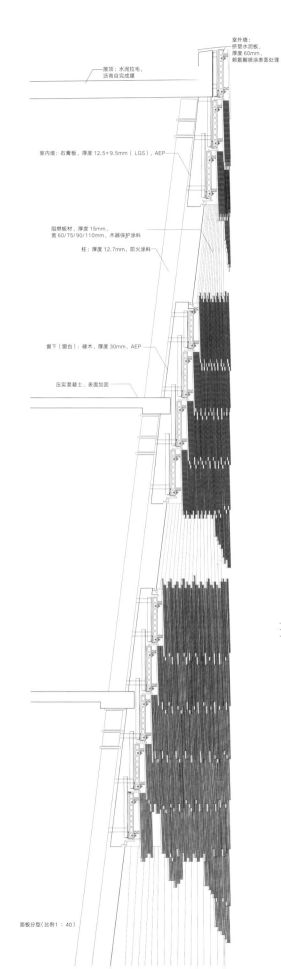

屋顶：水泥拉毛，
沥青自完成膜

室外墙：
挤塑水泥板，
厚度 60mm，
赖氨酸喷涂表面处理

室内墙：石膏板，厚度 12.5+9.5mm（LGS），AEP

阻燃板材，厚度 15mm，
宽 60/75/90/110mm，木器保护涂料

柱：厚度 12.7mm，防火涂料

窗下（窗台）：硬木，厚度 30mm，AEP

压实混凝土，表面加固

粒子化
particlizing

包装
wrapping

网格
grid

面板分型（比例 1：40）

"看步"品牌
蒙特拿破仑大街店
Camper Monte Napoleone

位置：意大利米兰

设计时间：2013.12—2014.06

建造时间：2014.08

主要用途：零售店

总建筑面积：88m²

结构顾问：大野日本（Ohno Japan）

建筑施工：Marukou建设（Marukou Construction）

14mm厚的结构用落叶松胶合板嵌合起来，便构成了一个立体格子状的空间。一个单元变换组合，就能生成货架、柜台、沙发和天花板等各种空间元素。超越传统中对变化的定义，更自由地探索身体与大地之间的联系，这是西班牙鞋履品牌"看步"的哲学，也是我们构思这个灵活的内部空间系统的灵感来源。

板材的厚度只有14mm，我们还是设法在货架面的边缘装了照明。电线通过嵌合部分从一块板连接到另一块板，排线是完全隐蔽的。杰出的意大利灯光设计师马里奥·南尼是我们的合作伙伴，没有他对工艺的执着追求，我们就无法实现这个极简的细节。

We created a three-dimensional lattice space by fitting together 14-mm-thick structural plywood made of karamatsu (larch). One unit, fitted together in different ways, generated all spatial elements, including the shelves, counters, sofas, and ceiling. I conceived this flexible interior system based on Camper's philosophy. It calls for discovering contact points between the body and the earth more freely by transcending conventional definitions of change.

Though the thickness was restricted to 14 mm, we managed to affix under-shelf lighting to the edges of the shelves. To achieve this, we ran the wiring from panel to panel through the fitting parts of the shelves. Thus, all wiring was entirely concealed, as was all plumbing. The extraordinary Italian lighting designer Mario Nanni, one of our frequent collaborators, worked with us on this design. Mario started his career as a craftsman. Without his extraordinary dedication to craftsmanship, we wouldn't have been able to realize these minimalistic details.

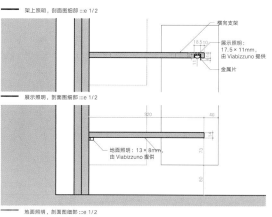

架上照明，剖面图细部 :::e 1/2

横向支架

18.510

展示照明：
17.5 × 11mm，
由 Viabizzuno 提供

金属片

展示照明，剖面图细部 :::e 1/2

320   40

地面照明：13 × 8mm，
由 Viabizzuno 提供

73

80

地面照明，剖面图细部 :::e 1/2

编织
weaving

堆叠
stacking

网格
grid

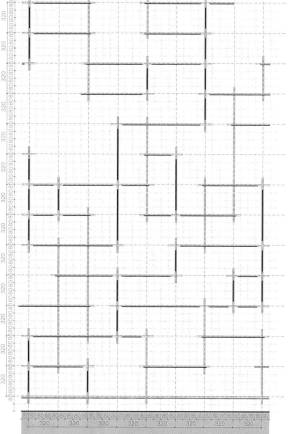

支架立面图示例

台北白石画廊
Whitestone Gallery Taipei

位置：中国台湾台北市
设计时间：2016.05—2016.10
建造时间：2016.11—2017.04
主要用途：画廊
总建筑面积：455m^2
结构顾问：江尻建筑结构设计事务所
建筑施工：RICER国际（RICER INTERNATIONAL）

将规格105×105mm的桧木材层层叠起，便在大厦的一楼空间里插入一个新的画廊。线形材料层层递进，形成一个个曲面，从外立面延伸到室内空间。柜台、展架都用同规格的木材构成，这是我们的"细胞法"的一次实践。

We transformed the first-floor space of an existing building into a new art gallery by inserting shifting stacks of hinoki (Japanese cypress) members, 105 mm square in section. The gradually shifting, linear members generate curved surfaces that extend from the facade to the gallery interior. The design is a variation on our "cellular method;" everything is constructed from a single type of squared timber, including the counters and shelves.

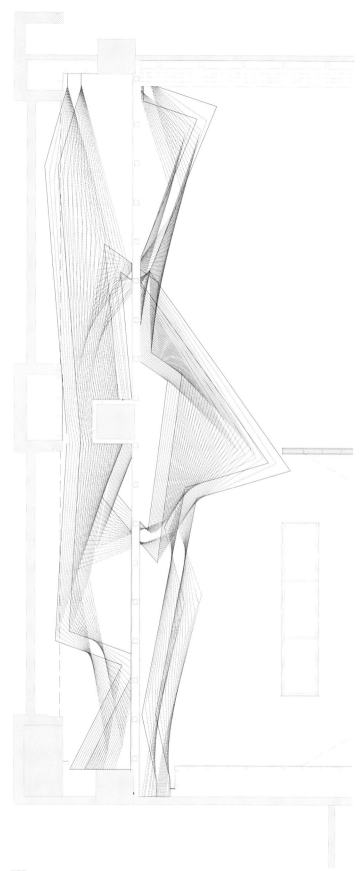

平面图

堆叠
stacking

多边形
polygon

洛桑联邦理工学院
艺术实验室
EPFL ArtLab

位置：瑞士洛桑
设计时间：2013—2015
建造时间：2013—2016.11
主要用途：艺术和科学展览馆、技术与信息展览馆、咖啡店
占地面积：13,500m²
总建筑面积：3,360m²
建筑施工：马蒂建设SA（Marti Construction SA）

这是洛桑联邦理工学院为融合艺术和科技新建的校园主楼。我们用穿孔钢板将集成木材夹在一起，做成柱子和梁。通过调整钢板和木材的厚度，使不同跨度框架的横截面的宽度都统一为120mm。细长的框架支撑起屋顶和屋檐，得到通透、灵活的建筑空间。这种钢与木的混合结构后来还被用在新品川车站和日本新国家体育馆的设计中。

根据建筑各个位置不同的边界条件，屋檐有时上扬，有时下沉，呈现多种面貌。不依赖墙面，而依靠屋檐的表现力来展现建筑的多种面貌，这是我们的一种新尝试。

This is the new main building of a campus created to fuse art and technology at the Federal Institute of Technology in Lausanne. We made columns and beams by sandwiching together perforated steel plates and laminated timber. Though spans differ throughout the building, we standardized the profiles of the structural frames at 120 mm by adjusting the thickness of the steel plates relative to the timber. The result is a building with transparent, flexible space, with a roof and eaves supported by slender frames. (Mixed structures of steel and wood also figure in our designs for the New Shinagawa Station and New National Stadium.)

The eaves of the ArtLab building respond to different boundary conditions; they rise up in certain places and dip down in others, giving the building a multitude of expressions. We were experimenting with a new technique of using eaves instead of walls to lend variety to the look of a building.

截面图（比例1：750）   Montreux爵士乐数字项目   艺术科学庭

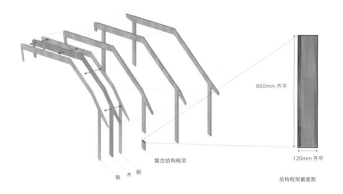

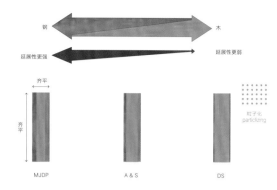

660mm 齐平

复合结构框架

钢 木 钢

120mm 齐平

结构框架截面图

钢 木

延展性更强 延展性更弱

齐平

齐平

粒子化
particlizing

MJDP

A & S

DS

所有立柱的横截面宽统一为 120mm，在这个前提下对集成木材和钢板的厚度进行调整，以符合结构的要求。全长 270m 的线形建筑中，存在多种剖面形态，从而为人们提供多样的空间体验。

All pillars have a uniform aspect width of 120 mm, and the thickness of the laminated wood members and steel plates were balanced to comply with the structural requirements. The linear structure which has a total length of 270 meters features a wide range of cross-sectional shapes in order to facilitate a diverse spatial experience.

包裹
wrapping

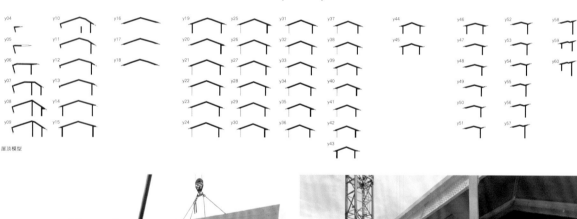

y04 y05 y06 y07 y08 y09
y10 y11 y12 y13 y14 y15
y16 y17 y18
y19 y20 y21 y22 y23 y24
y25 y26 y27 y28 y29 y30
y31 y32 y33 y34 y35 y36
y37 y38 y39 y40 y41 y42 y43
y44 y45
y46 y47 y48 y49 y50 y51
y52 y53 y54 y55 y56 y57
y58 y59 y60

屋顶模型

砌框
grid

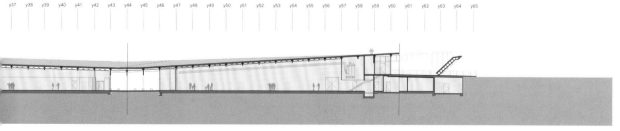

y37 y38 y39 y40 y41 y42 y43 y44 y45 y46 y47 y48 y49 y50 y51 y52 y53 y54 y55 y56 y57 y58 y59 y60 y61 y62 y63 y64 y65

数据广场

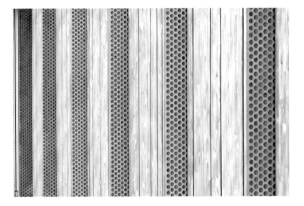

粒子化
particlizing

包装
wrapping

网格
grid

勃朗峰大本营
Mont-Blanc Base Camp

位置：法国霞慕尼
设计时间：2012.04—2014.07
建造时间：2014.07—2016.01
主要用途：总部、办公空间
占地面积：1,208m²
总建筑面积：2,500m²
机械及混凝土结构顾问：EGIS Grand EST公司（EGIS Grand EST）
结构顾问：巴尔特（BARTHES）
幕墙顾问：AR-C工作室（AR-C）

在法国霞慕尼一处可以望见勃朗峰的山坡上，我们用树木做了一个度假型的办公建筑群。建筑有着与山坡同样斜度的大屋顶。

建筑用宽600mm、厚30mm的橡木板覆盖外墙与屋顶。保留着树皮的木材富有野性，让建筑与阿尔卑斯山崇高的自然达到一种和谐。

On a sloping site surrounded by trees, with direct views of France's Mont Blanc and Chamonix, we created a resort-like office building with a large roof pitched at the same angle as the site.

The perimeter walls and roof of the building are covered with pieces of live edge oak, 600 mm wide and 30 mm thick. Our aim in using this wild material was to achieve balance between the building and the sublime natural environment of the Alps.

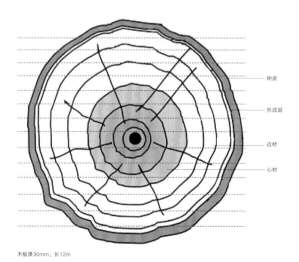

树皮
形成层
边材
心材

木板厚30mm，长12m

圣保罗日本屋
Japan House São Paulo

位置：巴西圣保罗
设计时间：2015.07—2016.06
建造时间：2016.06—2017.02
主要用途：多功能厅、多媒体空间、咖啡店、
商店、餐厅、画廊、会议室、办公室
总建筑面积：2,244.03m²
合作建筑：FGMF事务所（FGMF）
结构顾问：江尻建筑结构设计事务所
建筑施工：户田建设巴西分公司
（CONSTRUTORA TODA DO BRASIL）

在原有银行大厦的外立面上添加了新的木
构立面，大厦形象焕然一新。
A new wood facade was added to an
existing bank building to give the building
a new image.

新建的巴西日本文化中心——日本屋，坐落在圣保罗的繁华大街保利斯塔大道上。以不同角度随机横竖组合起来的桧木板材，像是在都市中营造出了一片形态丰富的森林。纵向材料规格为30×165mm，横向材料规格为30×150mm。入口的大跨度部分用碳纤维增加强度，下方以强度更大的巴西本地木材组合而成，在材料上也实现了文化交流。

This building, the new center of Japanese culture in Brazil, is situated along Paulista Avenue, one of the main streets of São Paulo. The vertical and horizontal members are planks of hinoki (Japanese cypress) with sectional dimensions of 30 mm by 165 mm and 30 mm by 150 mm, respectively. We tilted the planks at different angles and arranged them in a seemingly random pattern to create a rich expression, like a forest sprang up in the middle of the city. The materials we used served as a kind of cultural exchange: the long span portion at the entrance is reinforced with carbon fiber and the lower parts are made with a combination of high-strength Brazilian woods.

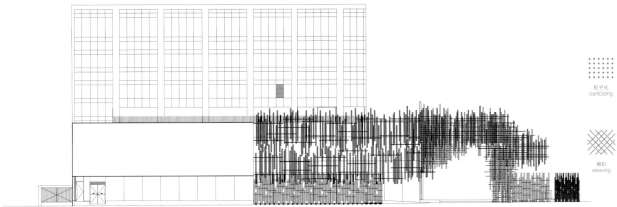

立面图(比例1：2500)

粒子化
particlizing

编织
weaving

包装
wrapping

网格
grid

Neowa圆顶
Neowa Dome

位置：韩国光州
时间：2016.10
主要用途：装置
总建筑面积：28m²
建筑施工：SY木工（SY wood）

我们用直径6mm的硬木榫，把18mm厚的结构用桦木胶合板拼接起来，组成一个圆顶装置。放弃嵌合的方式而改用木榫来连接，使得各个连接处的角度可以自由调整。此外，板材与板材之间是点状连接，整体上会感觉很轻盈。neowa在韩语中的意思是"斧头劈出来的木板"，日语对应的是"柿"（kokera，建造房顶时用的薄木板），英语对应的是shingle（木瓦）。韩国的木建筑比日本的木建筑更有野性，neowa一词正是这种意味的象征。

We constructed this dome using 18-mm-thick pieces of structural birch plywood connected with hardwood dowels, 6 mm in diameter. Joining the pieces with dowels instead of fitting them directly to each other allowed the connection angles of the joints to be freely determined and gave the entire structure a lightweight appearance. "Neowa," a Korean term meaning "board split with an ax," corresponds to "kokera" in Japanese and "shingle" in English. We used it here to symbolize the sense of roughness that typifies Korean wood architecture in comparison with Japanese wood architecture.

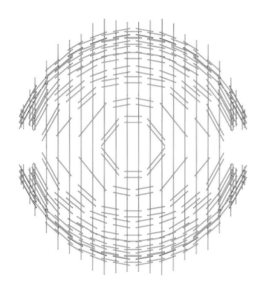

为了能从不同的视角获得通透或厚重的观感，我们确定了这样一个几何形态。
The geometry was determined with the objective of achieving an ambiguity that makes the dome feel both transparent and solid depending upon the point of view.

截面图和平面图（比例1：100）

粒子化
particlizing

编织
weaving

堆叠
stacking

网格
grid

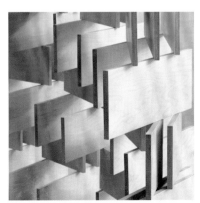

木灵（阿尔·塞拉装置）
KODAMA (Arte Sella Pavilion)

位置：意大利阿尔·塞拉

设计时间：2017.05—2017.12

建造时间：2018.03—2018.05

主要用途：装置

总建筑面积：约20m²

合作者：米兰理工大学 [马尔科·因佩拉多里（Marco Imperadori）、安德烈亚·瓦诺西（Andrea Vanossi）、费德里卡·布鲁内（Federica Brunone）、费德里卡·亚切里尼（Federica Iachelini）、克拉拉·里纳尔迪（Clara Rinaldi）]

结构顾问：D3木工（D3WOOD）、佐藤淳构造设计事务所

建筑施工：Ri-legno Srl事务所（Ri-legno Srl）

我们曾在木材上做出缺口，然后将它们组装成一个整体，并将这种做法用在室内设计上（如"看步"品牌蒙特拿破仑大街店和"Neowa圆顶"）。这次，我们计划用这个方法在森林里做一个永久性的装置作品。我们与意大利的木工及米兰理工大学的马尔科·因佩拉多里等人合作，用58mm厚的落叶松木片做了一个户外装置。

想象在意大利北部阿尔·塞拉的森林中，出现了一个树木的精灵——受此启发，我们做了一个球状的几何体。实际上只要改变缺口的组合方式，是可以组合成任何形态的。

Previously, we took the notched wood method used for creating pavilions and adapted it to interior design, in projects such as Camper Monte Napoleone and Neowa Dome. In this project, collaborating with Italian carpenters and Professor Marco Imperadori of the Polytechnic University of Milan, we once again adapted the method, this time to create a permanent pavilion in the forest. The outdoor structural system we designed uses solid, 58-mm-thick pieces of karamatsu (larch).

I imagined a spirit emerging from the forest in Arte Sella, Northern Italy and, inspired by this, I selected a spherical geometry. However, by modifying the notching and arrangement of members, the system can be used to produce any form.

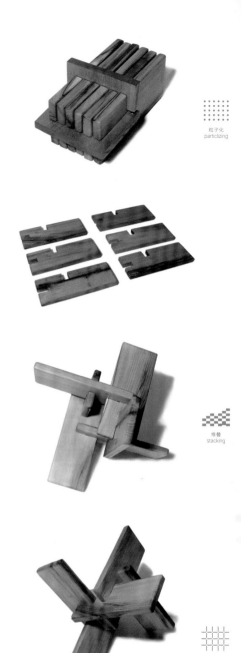

粒子化
particlizing

堆叠
stacking

网格
grid

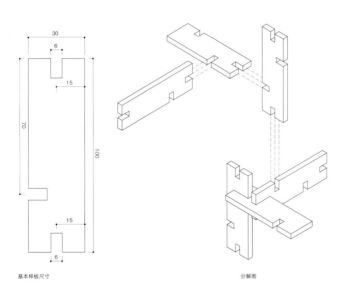

基本样板尺寸   分解图

Coeda House 咖啡厅
Coeda House

位置：日本静冈县
设计时间：2016.07—2017.02
建造时间：2017.04—2017.09
主要用途：咖啡厅
总建筑面积：141.61m²
建筑施工：桐山建筑（Kiriyama）

无数根80×80mm规格的罗汉柏木条，组合成一个通透、开放的伞状结构。它就像分出无数的细小树枝的粗壮树干，我们想表达的正是这种树木的美。最长的木材有12m长，为了避免变形，我们首先选取100mm规格的木料，然后干燥、切割，最后制成80×80mm尺寸的木材。

我们用90×90mm的实心钢柱支撑向外围伸展出去的屋檐般的结构，并在中央的"树干"里加入了碳纤维的悬吊装置，最终获得了一个与太平洋融为一体的开放空间。

By assembling slender sticks of hiba (Japanese cypress), 80 mm square in section, we created an open, transparent pavilion with an umbrella-like structure. The structure was inspired by the beautiful way in which the trunk of a thick tree branches off into countless twigs. Due to the fact that the members have a maximum length of 12 meters, the 100 mm members were milled, dried and cut to achieve a 80 mm square cross section to prevent warpage.

To help support the vertical load of the long, projecting eaves, we erected solid steel posts, 90 mm square in section, around the perimeter of the building and incorporated carbon fiber suspension members into the central trunk area. The result is a transparent space that seems one with the Pacific Ocean.

结构分析模型

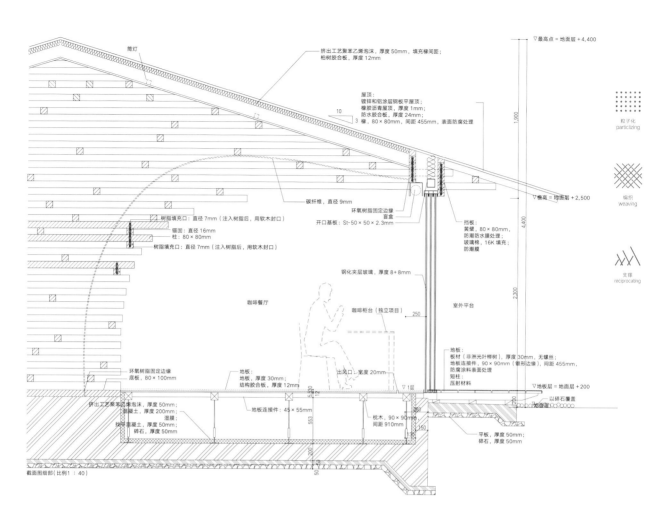

筒灯

挤出工艺聚苯乙烯泡沫，厚度50mm，填充椽间距；
柏树胶合板，厚度12mm

屋顶：
镀锌和铝涂层钢板平屋顶；
橡胶沥青屋顶，厚度1mm；
防水胶合板，厚度24mm；
3 椽，80×80mm，间距455mm，表面防腐处理

碳纤维，直径9mm

树脂填充口：直径7mm（注入树脂后，用软木封口）
锚固：直径16mm
柱：80×80mm
树脂填充口：直径7mm（注入树脂后，用软木封口）

环氧树脂固定边缘
盲盒

开口基板：St- 50×50×2.3mm

钢化夹层玻璃，厚度8+8mm

咖啡餐厅

咖啡柜台（独立项目）

出风口，宽度20mm

挡板
黄檗，80×80mm，
防潮防水膜处理；
玻璃棉，16K 填充；
防潮膜

室外平台

地板：
板材（非洲光叶榉树），厚度30mm，无螺丝；
地板连接件，90×90mm（锥形边缘），间距455mm，
防腐涂料表面处理；
短柱；
压射材料

环氧树脂固定边缘
底板，80×100mm

挤出工艺聚苯乙烯泡沫，厚度50mm；
混凝土，厚度200mm；
湿膜；
找平层混凝土，厚度50mm；
碎石，厚度50mm

地板，厚度30mm；
结构胶合板，厚度12mm

地板连接件：45×55mm

枕木，90×90mm
间距910mm

平板，厚度50mm；
碎石，厚度50mm

以碎石覆盖

截面图细部（比例1：40）

▽最高点 = 地面层 +4,400

▽椽高 = 地面层 +2,500

▽1层

▽地板层 = 地面层 +200

粒子化
particlizing

编织
weaving

支撑
reciprocating

网格
grid

汤布院漫画艺术博物馆
Comico Art Museum Yufuin

位置：日本大分县

时间：2014.02—2017.08

主要用途：博物馆

占地面积：1,716.64m²

总建筑面积：999.40m²

结构顾问：奥雅纳工程顾问公司（日本）
（Ove Arup & Partners Japan）

建筑施工：佐伯建设株式会社（SAIKI KENSETSU）

我们用5种不同截面的火烧杉木材，做出一种有着独特阴影的外立面，它随着季节、时间的变化呈现不同的样子。火烧杉木材浸泡过木材保护剂，表面的稳定性得到加强，触摸也不会弄脏手。

In this project, we used panels with 5 different sectional shapes to create facades that have distinctive shadows and change in appearance, depending on the season and time of day. The panels are made of yakisugi, Japanese cedar charred to improve durability. By soaking the yakisugi panels in a protective coating solution, we achieved a strong finish that isn't sooty to the touch.

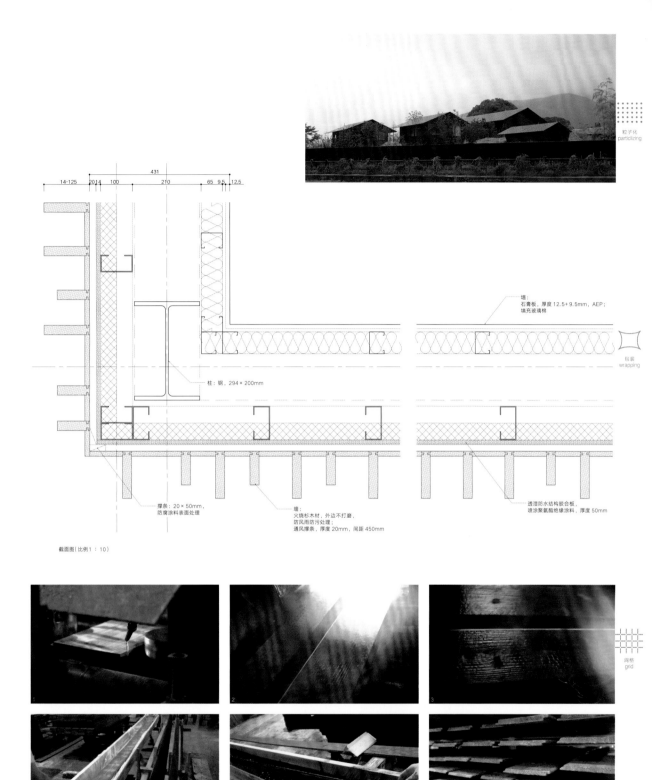

粒子化
particlizing

包装
wrapping

431

14-125　20 14　100　210　65 9.5 12.5

墙：
石膏板，厚度12.5+9.5mm，AEP；
填充玻璃棉

柱：钢，294×200mm

撑条：20×50mm，
防腐涂料表面处理

墙：
火烧杉木材，外边不打磨，
防风雨防污处理；
通风撑条，厚度20mm，间距450mm

透湿防水结构胶合板，
喷涂聚氨酯绝缘涂料，厚度50mm

截面图（比例 1：10）

网格
grid

碰触火烧杉木材手上会留下煤烟，用 Xyladecor 牌木材保护剂浸泡可以大大改善。
Since there is the chance of getting charcoal on your hands when you touch yakisugi, the yakisugi members were dipped in a Xyladecor solution in order to minimize the adherence of charcoal.

达令交流中心
The Darling Exchange

位置：澳大利亚悉尼
估计建造时间：2019
主要用途：商场、公共图书馆、儿童看护中心、餐厅
占地面积：1,116m²
总建筑面积：6,604m²

这座用于居民交流、福利服务和旅游的多功能建筑，位于悉尼老城区达令港的中心地带，内设商场、公共图书馆、儿童看护中心和餐厅。

将厚30mm、宽95~195mm规格的固雅木（乙酰化处理的新西兰松木）弯折成条带状，缠绕着建筑螺旋上升，从而把大地与天空、陆地与海洋连接起来。调整木条与垂直钢管之间的距离，可以让条带呈现的纹理富有自然的随机性。

This multi-use facility for community exchange, welfare services, and tourism will be situated at the heart of Darling Harbor, Sydney's old city center. It will contain a market, library, child care center, and restaurants.

Wood ribbons will wrap the building in a spiral formation, reconnecting the earth and sky and the land and sea. We created these ribbons by bending and joining together 30-mm-thick pieces of Accoya (acetylated New Zealand pine) that range in width from 95 mm to 195 mm. The stripes of the Accoya cladding were provided with a natural randomness by adjusting the distance that they protrude from the vertical steel pipes.

编织
weaving

包装
wrapping

螺旋
spiral

工厂加工外观

+

工厂加工内部

+

现场加工外观

+

现场加工内部

=

组合层次

纸

Paper

纸是用树木的纤维制成的，所以纸与木可以归为一组材料。但树木不断纳入自己死去的组织，变得牢固、强硬，而纸没有这样的性质，总是柔软的。

柔软的性质对于建筑材料来说一般是缺点，但是如果我想在整体上获得一种松弛的感觉，那么柔软就会成为一种武器。所以，每当我想让建筑变得更加柔软、松弛时，纸就有了用武之地。

纸的有趣之处在于，它可以作为液体存在。制作和纸时，首先是分解构树和桑树等树木的纤维；然后浸泡在黄蜀葵煮出来的黏液中，成为液态的纸浆；接着用竹子做的细筛网抄起，晾干即可。如果把液态的纸浆涂到别的物体上面，晾干后，这个物体就像是和纸做出来的了。涂在金属上，冷硬的金属表面就会拥有和纸的朦胧感和柔软。

圣保罗的日本屋（2017）、巴黎的安东尼·克拉夫档案馆（2017），都是把金属网浸入纸浆再晾干，变成"纸质"的。

和纸的朦胧感很大程度上来自黄蜀葵。黄蜀葵的作用并不是像胶水一样把构树或桑树等树木的纤维黏合在一起，而是让这些纤维均匀地分散在黏液里，不会纠结成团。纤维不是由胶水黏合在一起的，而是自由地悬浮在黏液里。等到干了，液态的纸浆就变成了固态的和纸。液态的东西不知不觉变成了固体，因为这种优雅的转变，

Paper is made from wood fibers, so it can be placed in the same group as wood. Wood becomes harder and stronger by incorporating its own dead cells, but paper does not have this quality, so it is always soft and supple.

Suppleness can be regarded as a defect in an architecture material, but for our purpose of achieving overall flexibility it becomes an advantage. When the goal is to make the entire project looser and more flexible, more applications will be found for paper.

An interesting thing about paper is that it can exist in a liquid state. The process of making washi starts with separating the fibers of mitsumata or mulberry shrubs and soaking them in a starchy solution made by boiling the roots of the tororo aoi plant. The pulpy liquid thus prepared is scooped up with a filter called a takesu—a fine-mesh screen made from bamboo—and then dried. The result is washi. But the liquid can also be painted onto other objects and dried to give them a paper coat. When painted on metal, it can transform a cold, hard surface into a soft one with the vague texture of washi.

At the Japan House São Paulo (2017) and the Archives Antoni Clavé (2017) in Paris, we soaked expanded aluminum mesh forms in the pulpy liquid and then dried them to transform them into washi. The vagueness of washi is due in large part to the effect of the tororo aoi, which acts less as a glue to bind the mitsumata or mulberry fibers and more to keep them aligned without becoming tangled. All that remains left to do is to let it dry. The fibers are not bonded to each other with glue, but float freely in the solution. After a while, the liquid has evaporated and the paper

1

我想把和纸称为"冻结的液体"。正是黄蜀葵造就了和纸的液体性。对于全部由固体构成的建筑物来说，和纸拥有的"液体性"非常珍贵，因此日本的传统建筑对纸这种材料非常重视。在日本，无论是城市还是单独的建筑，它们与水的关系总是设计方面的决定性因素：建筑的形态取决于如何在雨天保护好建筑，庭园的构成以水为中心，建筑材料的选择也围绕着水来考虑。

has become washi in its solid form. The elegance of a liquid that becomes a solid, seemingly while no one is watching, invites the description "crystallized liquid." The tororo aoi also contributes to the liquid nature of paper. In architecture composed entirely of solids, the liquidity that continues to exist in washi is an invaluable asset. This is why traditional Japanese architecture emphasized paper. In Japan, the relationship with water was determinative for all urban and architectural designs. The shape of the architecture was determined by the need to protect it against rain. Gardens were centered around water, and water was decisive in the selection of architectural materials.

2

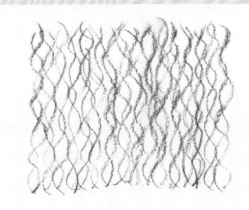

3

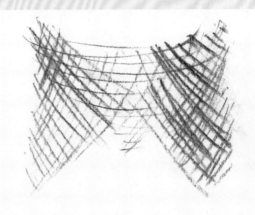

4

Sketches by Kengo Kuma
1. Takayanagi Community Center
2. Paper Snake
3. Paper Cocoon
4. Seigaiha

隈研吾手稿
1. 高柳町地域交流中心·阳光之家
2. 纸蛇
3. 纸茧
4. 青海波

高柳町地域交流中心·
阳光之家
Takayanagi Community Center

位置：日本新潟县

设计时间：1998.02—1999.07

建造时间：1999.11—2000.04

主要用途：活动中心

总建筑面积：87m²

结构顾问：中田捷夫及合伙人

建筑施工：长井公司（Nagai cooperation）

合作建筑：栗原胡藤（Hutoshi Kurihara）、

高桥裕美（Hiromi Takahashi）、

小林保雄（Yasuo Kobayashi）

新潟县柏崎市高柳町以"门出和纸"闻名，我们与和纸匠人小林康生先生合作，一起做了这座通体透光的纸建筑。在日本，普通住宅中使用玻璃是明治中期以后的事，从前都是靠纸门窗和木质的挡雨板来抵御风雨的。于是我们想重现一座朴素、坚固的传统式样的日本房屋。

我们给和纸刷上柿漆（柿子发酵而成的防腐剂，含单宁）和魔芋胶，提高和纸的防水性能。地板、梁和柱都用这种处理过的和纸包裹起来，营造了一种像在蚕茧里一样的温和空间。

We designed this building in collaboration with Yasuo Kobayashi, the washi craftsman in charge of Kadoide Washi. It is situated in Takayanagi-cho, a village in Kashiwazaki, Niigata that is famous as the headquarters of Kadoide Washi. Glass has only been used in residential architecture in Japan since the middle of the Meiji period. Before that, sliding paper screens (shoji) and wooden shutters (amado) were all that kept out the wind and rain.

With this in mind, we decided to create a simple, sturdy house using only traditional technologies. We waterproofed washi by coating it with persimmon tannin and konnyaku, a substance made from konjac corms. By completely wrapping the floor, columns, and beams with this specially treated washi, we created a gentle space that feels like the inside of a cocoon.

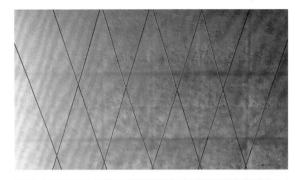

包装
wrapping

为了给预制混凝土施加拉力，将直径 5mm 的钢丝像翻花绳那样交织起来，形成了一种非常通透的支撑结构。结构设计由中田捷夫担任。

In order to give tensile force to precast concrete, steel wire with a diameter of 5 mm was woven like a string figure to achieve a brace structure with a high level of transparency. Structural engineer was by Katsuo Nakata.

网格
grid

纸蛇
Paper Snake

位置：韩国安养市

时间：2005

主要用途：装置

占地面积：42m^2

结构顾问：江尻建筑结构设计事务所

我们用多层胶合的牛皮纸做成40mm厚的蜂窝板，夹在两层3mm厚的FRP（纤维增强复合材料）板中间，涂上环氧树脂胶，再加压，使之成为轻量透明又具有足够强度的结构板。加压借用的是一台制造自动售货机用的压力机。这些板材的面并不相互平行，而是错位连接起来，形成一个螺旋状的稳定结构体。这个轻盈的流线型装置就像一条蜿蜒穿过森林的蛇。

First, we took 40-mm-thick "paper honeycombs" made with multiple layers of glued craft paper and sandwiched them between 3-mm-thick FRP panels. Next, we coated the "sandwiches" with epoxy adhesive and pressed them using a press for manufacturing vending machines. This produced highstrength structural panels that are lightweight and transparent. Rather than positioning the faces of the panels in parallel with each other, we rotated them and connected their edges, creating a stable, spiralshaped structure. In this way, we realized a lightweight, linear shelter that meanders through the forest like a snake.

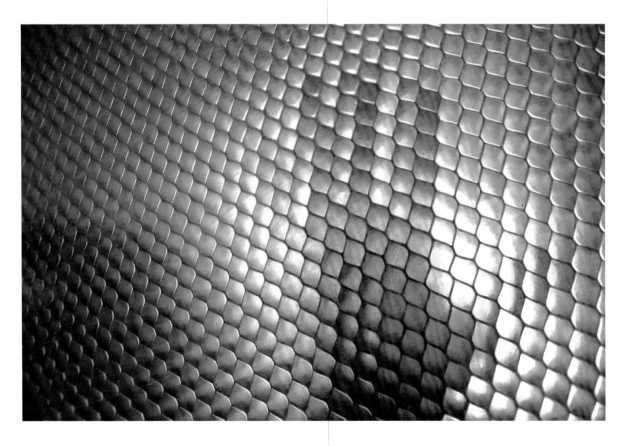

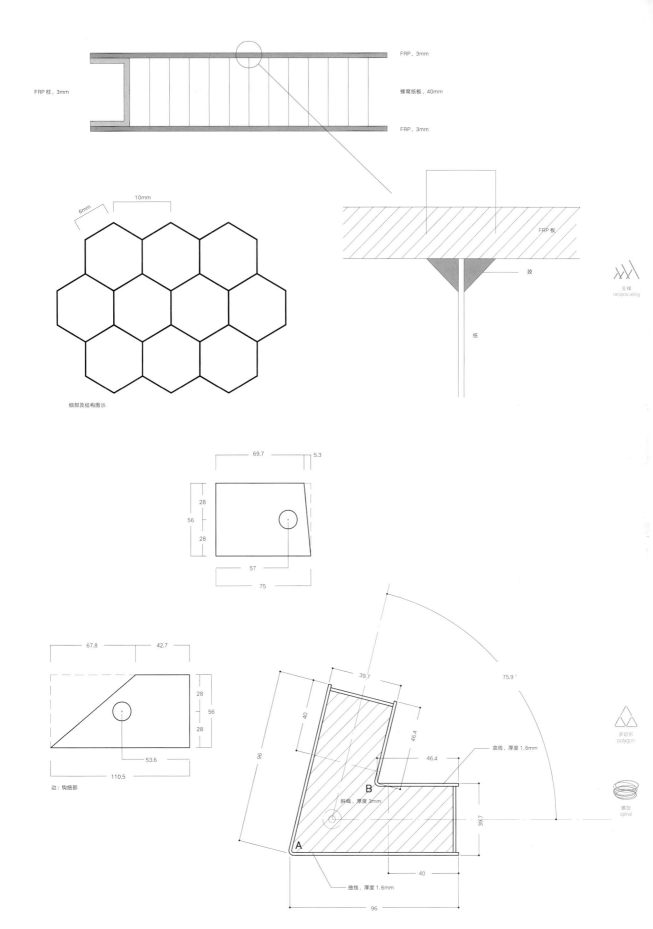

FRP，3mm

FRP 柱，3mm

蜂窝纸板，40mm

FRP，3mm

10mm

6mm

FRP 板

胶

纸

支撑
reciprocating

细部及结构图示

69.7   5.3

28

56

28

57

75

多边形
polygon

螺旋
spiral

67.8   42.7

28

56

28

53.6

110.5

边：钩细部

39.7

40

46.4

46.4

96

曲线，厚度 1.6mm

75.9°

B

斜线，厚度 3mm

39.7

A

40

曲线，厚度 1.6mm

96

薄弱的角落部分采用钢构件,钢板同样夹在 FRP 板中。纸蜂窝与 FRP 板之间需要用到透明的黏合剂,这方面我们得到了日本施敏打硬株式会社的协助。

Corner materials made using steel plates were provided on the corners which are the weak points, and sandwiched with FRP panels. We worked together with the Cemedine company since the paper honeycomb and FRP materials needed to be joined with a transparent adhesive.

纸砖
Paper Brick

位置：中国上海
设计时间：2014.04—2014.05
建造时间：2014.06
主要用途：装置
占地面积：480m²（展览空间）
合作建筑：艾敬（Ai Jing）
结构顾问：奥雅纳工程顾问公司（上海）
（ARUP Shanghai）

把废报纸用水浸泡溶解，灌入模具中。仅依靠纤维素分子之间的氢键作用力，纸浆在干燥后即可成形，成为性能优越且环保的纸砖。这是对鸡蛋包装盒技术的一次实验性应用。纸砖的凹凸部分相互咬合，垒砌起来，可以自由地构成一个富有艺术性的空间。这种凹凸咬合垒砌的结构手法，很容易让人想起乐高积木，在我们做的"水砖"系列（水砖屋）中也可以见到。

Using a process called dry molding that relies only on hydrogen bonding with paper, formed blocks by dissolving old newspapers in water, poured the solution into molds, and letting them dry. This process, based on a technique used to make egg cartons, enabled us to create interlocking paper blocks that are lightweight and environmentally friendly. We fit the blocks together into stacks and arranged the stacks into an art space. The interlocking block structure is of the same kind as Water Block and Water Branch House, not to mention Legos.

堆叠
stacking

网格
grid

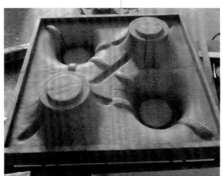

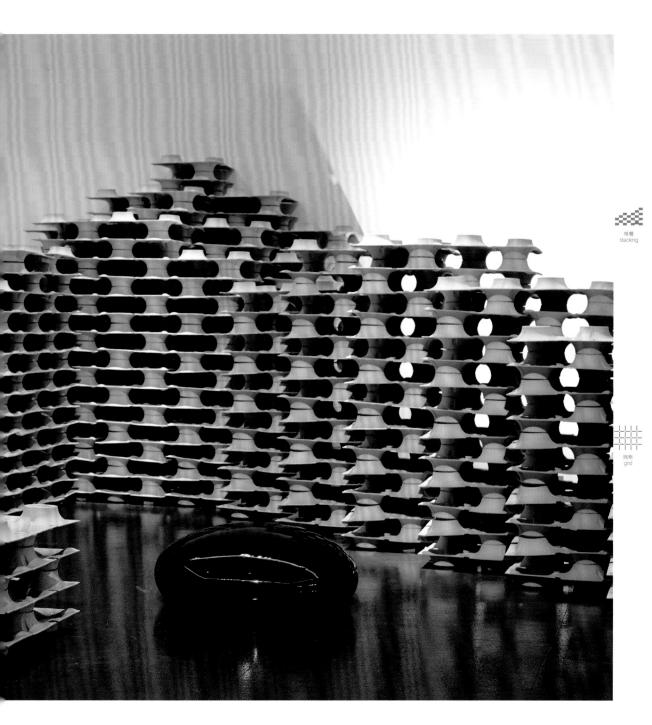

纸茧
Paper Cocoon

位置：意大利米兰
设计时间：2014.10—2015.06
建造时间：2015.04—2015.12
主要用途：装置
尺寸：纵深13.5m、宽5m、高2.5m
结构顾问：江尻建筑结构设计事务所

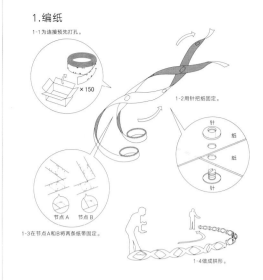

1.编纸
1-1为连接预先打孔。
×150
1-2用针把纸固定。
针
纸
纸
针
节点A 节点B
1-3在节点A和B将两条纸带固定。
1-4做成拱形。

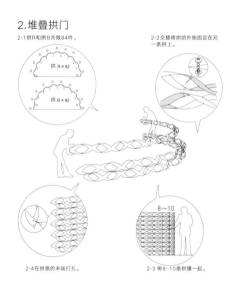

2.堆叠拱门
2-1拱R和拱B共做84件。
拱 R×42
拱 B×42
2-2交替将拱的外侧固定在另一条拱上。
2-4在拱条的末端打孔。
8~10
2-3 将8~10条拱摞在一起。

钢纸是一种特种纸，制作时将原纸浸在氯化锌溶液中，通过氯化锌对纤维素的膨润作用使原纸得到强化，1mm的厚度也能拥有很强的刚性。将钢纸卷起来能够进一步增强刚性。基于此，我们用钢纸卷交织成一个高2.5m、宽5m、纵深13.5m的轻巧的隧道状空间。

Vulcanized paper is a special type of paper that has been soaked in a zinc chloride solution and strengthened through a swelling effect the zinc chloride has on its fibers. Even one-mm-thick paper of this type has a high stiffness. Rolling the paper into tubes and weaving them together to further increase the stiffness, we created a tunnel-shaped space that is 2.5 m height, 5 m width, and 13.5 m depth.

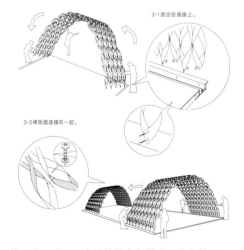

3.将连接在一起的拱立起来
3-1固定在底座上。
3-2将侧面连接在一起。

按照施工步骤说明，当地的学生们搭建了这个装置。
Manuals were distributed, and local students built the structure.

编织
weaving

支撑
reciprocating

包裹
wrapping

（按时间排列）1. 空运了 6 箱纸。/ 2 和 3. 150 根纸管被组合起来。/ 4 和 5. 一根纸管长 10m，铺满了整个工作现场。/ 6. 连好的拱门足够刚硬，可以单独立起来。

(Timeline on Right) 1. Six boxes of paper were transported by air. / 2&3. 150 paper tubes were combined. / 4&5. One paper tube was 10 meters long, requiring the entire site to perform work. / 6.The connected arch was rigid enough to stand by itself.

我们研究了用同材料的纸（图 1）和金属（图 2）固定，最终采用了塑料铆钉（图 3）。

Caps made from the same paper (1) and metal tabs (2) were evaluated, but in the end, the decision was made to use plastic caps (3) for packing purposes.

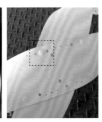

螺旋
spiral

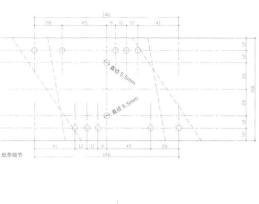

纸条细节

侧面连接处

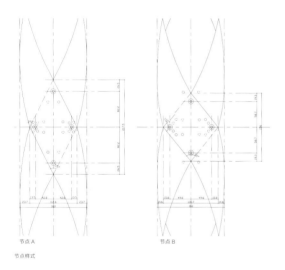

节点 A   节点 B

节点样式

每隔 150mm 开 12 个孔，使纸可以自由连接。

Twelve holes were made at an interval of 150 mm to allow connections to be freely made.

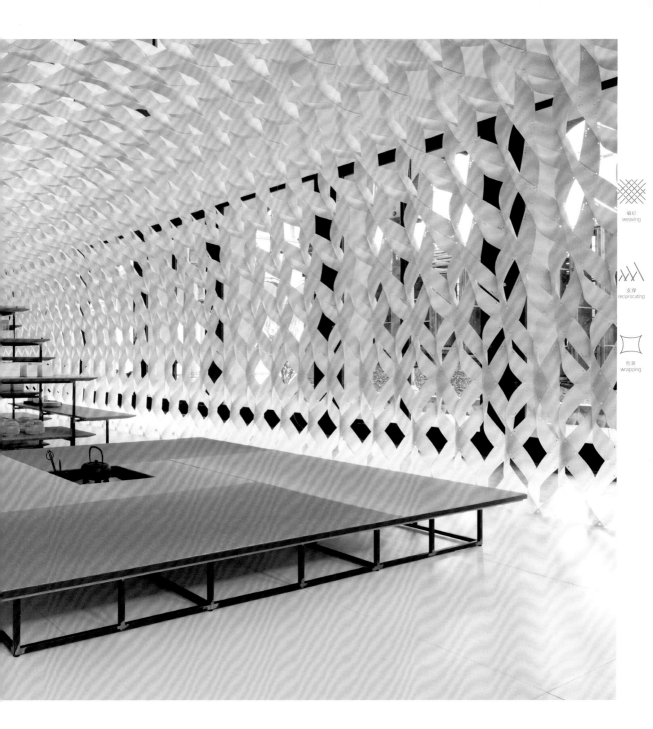

青海波
Seigaiha

位置: 任意
设计时间: 2005.01—2005.03
建造时间: 2005.03
主要用途: 装置（移动茶屋）
尺寸: 纵深2.6m、宽2.3m、高2m

在许多项目中，我们都尝试过在薄材上切缝，然后拉伸，将其做成轻盈透明的网。用金属板做出来的叫作"金属拉网"。这次我们做的是"和纸拉网"，用的是门出和纸的代表人物、和纸匠人小林康生先生以日本国产长纤维构树为原料制作的高强度手抄和纸。我们在和纸上做出切缝，将其悬挂起来，作为一个可移动的茶室。和纸材料很轻，可以装在一个卷筒里搬运。拉伸后出现的波浪形图案，很像日本传统图案中的"青海波"。

In many projects we have experimented with slitting and stretching thin surfaces to create lightweight, transparent meshes. When these operations are performed on metal panels, the resulting meshes are called "expanded metal." In this work, we used "expanded washi." The material is a sheet of high-strength washi made by Yasuo Kobayashi, the head washi craftsman at Kadoide Washi, using only domestically grown, long-fibered paper mulberries. We covered the sheet in slits and hung it to create a mobile tea room. "Expandable washi" is light enough to be carried around in a tube. I gave the project this name because a wave-form pattern resembling the traditional Japanese seigaiha pattern appears when the sheet is expanded.

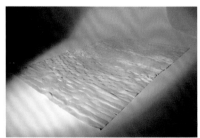

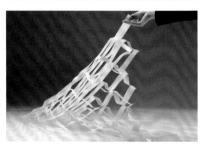

做好切缝的和纸可以装在很小的箱子里搬运，在现场拉开就能搭出一个茶室空间。

The washi paper that contains cuts can be transported in a small suitcase, allowing the material to be expanded on site to create a tea room.

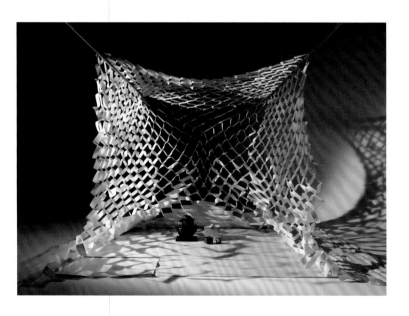

包装
wrapping

网格
grid

扩展计划

安东尼·克拉夫档案馆
Archives Antoni Clavé

位置：法国巴黎
时间：2015.04—2017.12
主要用途：办公室
占地面积：190m²
总建筑面积：240m²
结构顾问：诺曼德工作室（Atelier Normand）
内部工程师
建筑施工：诺曼德工作室

通过对干燥工艺进行调整，使和纸获得立体感。

A three-dimensional washi paper texture was achieved by fine tuning the drying technique.

我在和纸匠人小林康夫的工作室看到了一个和纸筐，是把和纸的纸浆涂在竹筐上做成的。这给我带来了启发。

安东尼·克拉夫（1913—2005）是20世纪西班牙杰出的艺术家，我们为他的档案馆所做的设计中就用上了涂有和纸纸浆的金属拉网。涂上和纸纸浆，既可以改变金属拉网的透明度，又可以调整顶部洒落光线的量和质感。在巴西圣保罗的日本屋中，我们也用了同样的设计。巴黎的这个项目由小林先生亲自在现场制作，而巴西的项目由当地的年轻人动手制作，这也是技术传承的一个尝试。

When visiting the atelier of Yasuo Kobayashi, I was inspired by washi mabushi baskets, bamboo baskets coated in the pulpy liquid used to make washi.

In our design for the archive of Antoni Clavé (1913—2005), one of Spain's leading 20th century artists, we adapted the washi mabushi technique by coating expanded aluminum panels in washi and using them to envelop the space. The technique enabled us to adjust the transparency of the expanded metal and control quantity and quality of light entering through skylights. Mr. Kobayashi himself visited the site to produce the panels. We also used this detail on the interior of Japan House São Paulo in Brazil. In that project, however, as an experiment in technology transfer, we had local youths coat the panels.

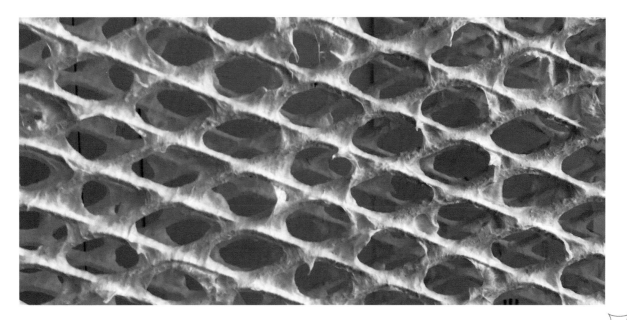

包装
wrapping

网格
grid

Earth

如果说和纸是一种介于液体和固体之间的存在，那么土就可以算是一种介于固体、液体与气体之间的物质。我老家的房子，土墙上布满了裂痕，榻榻米上总是散落着土屑。土屑与灰尘很难区分，空气中的灰尘与土屑乃至土墙总是浑然一体的。而土墙，用湿抹布擦一下，很容易就转化成液体，成为抹布上的污渍。土墙实在是一种变化自由的东西。

利用土的这种液体性，把混有玻璃纤维的土浆喷涂到金属上，冷冰冰的金属就变成了带有温度和湿度的土。我从泥灰匠人挟土秀平那里学到了这个技术。日本自古以来，常以水为媒介，将土以各种形式用在建筑上。

这其实是把土地令人安心的质感带入建筑这种人造物里去。很多人可能不知道，过去人们会在新铺的榻榻米上撒土，利用土将新的人造物的人工性抹去，变成亲和人体的东西。

土地上会长植物。从土地的性能上来说，日本的土很特别。因为有很多火山，日本的土中富含火山灰带来的营养物质，很容易长出杂草。放眼全世界，这样的土也是不多见的。我们做海外项目的时候，经常发现建筑周边很难长出杂草，建筑与周围环境的融合需要很长时间。

有一种陶瓷类的新型材料，叫作"多孔陶瓷板"（greenbiz），是将染布过程中产生的废渣进

If washi is an intermediate material between liquid and solid, then earth could be called an intermediate between solid, liquid, and gas. The earthen walls of the house where I grew up were full of cracks, and there was always a scattering of wall dust on the tatami f loor mats. The wall dust was hard to tell from airborne dust, so to me airborne dust and debris were on a continuum with our earthen walls. When the walls were wiped with a damp rag they would revert to water and leave a stain on the rag. Our earth walls were truly metamorphical, able to change at will.

When the liquid aspect of earth is used to apply earth mixed with glass fiber to metal, the cold metal is transformed into warm, moist earth. I learned this technique from the plaster craftsman Syuhei Hasado. Since long in the past, via water, earth has been used in various forms in Japanese architecture.

Architecture is artificial, and people wanted to bring in a reassuring touch of the solid earth underfoot. Although it is not well-known today, people used to sprinkle earth on new tatami mats. The earth would allay the artificiality of the new mats and make them more congenial for the body.

Plants and moss grow on the earth. Especially in Japan, where nutrient-rich volcanic soil is common, weeds will flourish on any spot of soil that is left unattended. This kind of soil seems to be rare. In overseas projects, weeds don't grow around the site, so that it takes time for the architecture to look like it belongs in the place.

There is a microporous ceramic material called

行干燥后做成的。我也把它算作土一类的材料。在这种轻量多孔质的板材上，植物可以直接扎根，长成一片茂密的绿色，因此它可以用来做绿化斜坡屋顶。土不仅仅是粉屑的集合体，它还是颗粒物，同时也是液体，能够孕育出生命。

greenbiz made by drying the slag produced when cloth is dyed. I consider it to be akin to earth. Plants can take root and flourish on sheets of this material. With greenbiz as a foundation, it is now possible for greenery to grow on thin, lightweight pitched roofs. Earth is not simply a collection of particles. In addition to particles, it is also a liquid that fosters life.

1

2

Sketches by Kengo Kuma
1. Adobe Repository for Buddha Statue
2. Mesh / Earth
3. Mushizuka (Mound for Insects)
4. Portland Japanese Garden Cultural Village

隈研吾手稿
1. 安养寺阿弥陀佛木坐像收藏设施
2. 网 / 土
3. 虫冢
4. 波特兰日本庭园文化中心

3

4

安养寺阿弥陀佛
木坐像收藏设施
Adobe Repository for
Buddha Statue

位置：日本山口县

设计时间：2001.02—2002.02

建造时间：2002.03—2002.10

主要用途：寺庙

占地面积：2,036.75m²

总建筑面积：63.23m²

结构顾问：中田结构设计事务所

建筑施工：冈崎建设（Okazaki Construction）

在这个项目中，我们重现了当地一种叫作"泥砖"的风干土砖的技术，据说这是由朝鲜传入日本的。风干土砖曾是中东到中国乃至欧亚大陆很常见的建筑技术，但日本只有这里使用过这种技术。这一带的土叫"丰浦土"，是一种颗粒均匀的优质土，被用作混凝土结构标准的参考及汽车发动机的研磨剂。此外，这种土还有很好的保温性能。我们就用这种土做成风干土砖，来建造这个12世纪木佛像的收藏设施。作为重要文物的收藏设施，不用空调，而是依赖建材的天然性能调节空气的温度和湿度，这是非常特殊的做法。

泥灰匠人久住章用当地的土和水和泥，混入稻秸制作了这批土砖。接触地面的部分添加了混凝土，以防因雨水浸泡倒塌。在需要通风的地方留出缝隙，辅以钢板加固。

In this project, we revived the use of adobe technology, which is said to have been introduced to Japan by Korean craftsmen. Although adobe was once the most common architectural technology on the Eurasian continent, from the Middle East to China, there are no examples of it in Japan outside this region. Soil from the region, known as "Toyoura soil," is a high-quality soil with evenly sized particles. It has been used as a reference in concrete structural standards and as polish for automotive engines. Adobe made with the soil has high heat retention properties. This enabled us to use the adobe to create a naturally ventilated building. (This is quite exceptional, considering the building is the storage facility is for a 12th-century wooden Buddha statue, designated an Important Cultural Property.)

Plasterwork craftsman Akira Kusumi made the adobe by dissolving the local soil in water and mixing in straw fibers. Cement was incorporated into blocks that touch the ground to prevent them from collapsing in the rain. In areas requiring ventilation, slits were left between blocks and reinforced with steel plates.

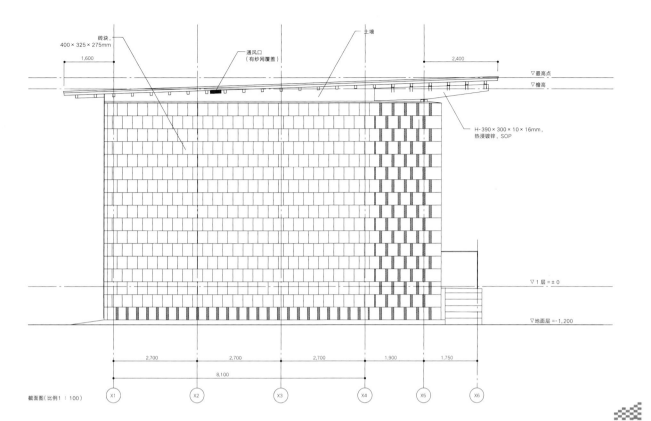

砖块，
400 × 325 × 275mm

土墙

通风口
（有纱网覆盖）

1,600

2,400

▽ 最高点

▽ 檐高

H-390 × 300 × 10 × 16mm，
热浸镀锌，SOP

▽ 1 层 = ± 0

▽ 地面层 =-1,200

2,700   2,700   2,700   1,900   1,750

8,100

截面图（比例 1：100）

X1   X2   X3   X4   X5   X6

堆叠
stacking

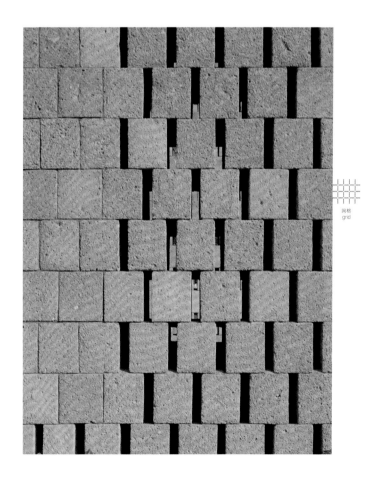

网格
grid

网 / 土
Mesh / Earth

位置：日本东京
设计时间：2009.11—2010.06
建造时间：2010.07—2011.02
主要用途：私人住宅
占地面积：158.11m²
总建筑面积：182.82m²
结构顾问：大野日本
建筑施工：松井建设株式会社

这是一所建在公园梅林中的住宅，我们尝试用涂了土的金属网把房子遮上，让它与梅林融为一体。用直径3.2mm的不锈钢丝编成100×100mm尺寸的网格，再喷上混有玻璃纤维的土，安装在外墙上，将建筑包裹起来。这样建筑就有了一个柔和的外观、模糊的轮廓，还有一定的重量感。

This house is in the middle of an ume (Japanese apricot) grove in a park. We tried to make it blend into the grove by covering it in soil-coated metal mesh. The mesh was woven from stainless steel members, 3.2 mm in diameter, based on a 100-mm-square grid. It was sprayed with a mixture of glass fiber and soil and, finally, it was wrapped around the building's perimeter walls. This enabled us to realize a soft-looking facade with an ambiguous outline and a sense of weight.

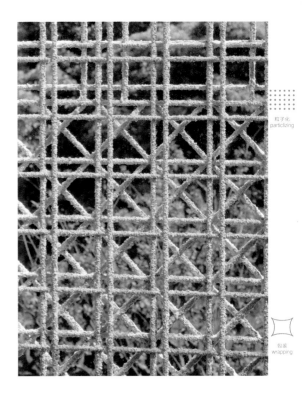

粒子化
particlizing

包裹
wrapping

喷涂土层的工作在飞弹高山的泥灰匠人挟土秀平的工作室完成，然后运到现场。

Spraying was performed at the workshop of Shuhei Hasado, plasterwork craftsman in Hida-Takayama, and the members were delivered to the site.

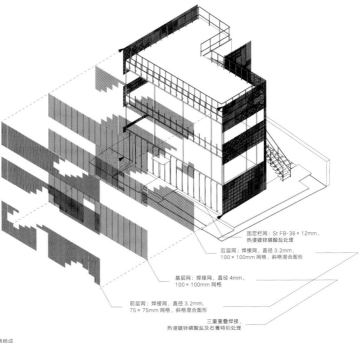

网格
grid

固定栏网: St FB-38×12mm,
热浸镀锌磷酸盐处理

后层网: 焊接网, 直径 3.2mm,
100×100mm 网格, 斜格混合图形

基层网: 焊接网, 直径 4mm,
100×100mm 网格

前层网: 焊接网, 直径 3.2mm,
75×75mm 网格, 斜格混合图形

三重重叠焊接,
热浸镀锌磷酸盐及石膏特别处理

石膏网格组成

虫冢
Mound for Insects

位置：日本神奈川县
设计时间：2013.06—2015.06
建造时间：2014.10—2015.06
主要用途：纪念性装置
占地面积：57m²
尺寸：直径8.5m，高2.5m
结构顾问：江尻建筑结构设计事务所

位于镰仓建长寺（1253年创建）内的"虫冢"是一个供养昆虫的纪念性装置，由神经学家兼昆虫爱好者养老孟司先生提议建造。从养虫的笼子得到灵感，我们把不锈钢网格的笼子以螺旋状堆叠起来。泥灰匠人挟土秀平给网格喷涂了由玻璃纤维、当地土壤以及黏合剂混合而成的土浆，使之成为一个透明而有质感的纪念性装置。

和其他涂层材料不同，土会随着时间变色，长出苔藓，最后融入大地。

This monument to insects was proposed by neuroscientist and bug enthusiast Dr. Takeshi Yoro. It stands on the grounds of Kencho-ji, a temple in Kamakura that was founded in 1253. Taking cues from insect cages, we designed a spiraling stack of cases made from stainless steel mesh. Plasterwork craftsman Shuhei Hasado sprayed the mesh with a mixture of glass fiber, adhesive, and local soil. The result is a monument that is transparent and also has texture.

Sprayed soil differs from other sprayed materials in that, over time, it will change colors, grow moss, and be reabsorbed into the ground.

喷涂当地的土。

Local clay was sprayed on.

从后部高墙上俯瞰的虫冢。中间放置着养老孟司先生收藏
的草履虫雕像。

Mushizuka viewed from rear wall. A paramecium objets d'art
owned by Dr. Takeshi Yoro was placed in the center.

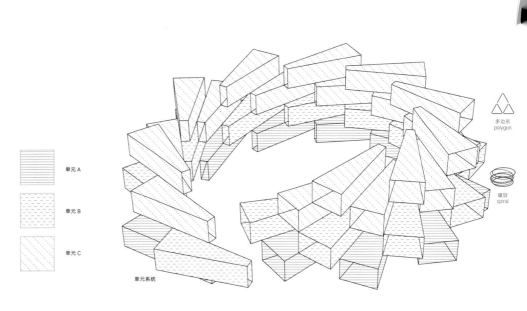

单元 A

单元 B

单元 C

单元系统

诺华上海园区多功能楼
Novartis Shanghai Campus Multifunction Building

位置：中国上海
设计时间：2009.11—2016.05
建造时间：2010.10—2016.05
主要用途：多功能建筑
总建筑面积：960m²
结构顾问：RFR公司（RFR）

我们在医药公司"诺华"的上海园区里做了一个有别于周围四四方方建筑的交流空间。空间被绿色覆盖，能够让人感受到泥土的气息。我们受到森林中自由生长的树木启发，用微微倾斜的木构件支撑多角的绿色屋顶。多个屋顶形成的微妙褶皱，以及植被具有的孔隙，给园区带来一种润泽的生机。

This communication space is situated at the center of a pharmaceutical company's R&D campus. I wanted to give it an earthy feel and cover it with greenery to distinguish it from the boxy office buildings surrounding it. Randomly arranged trees in a forest inspired a framework of subtly tilting wood members, which support green roofs pitched at multiple angles. The minute pleats formed by the many roof surfaces and the porosity of the vegetation bring a sense of moisture to the arid campus.

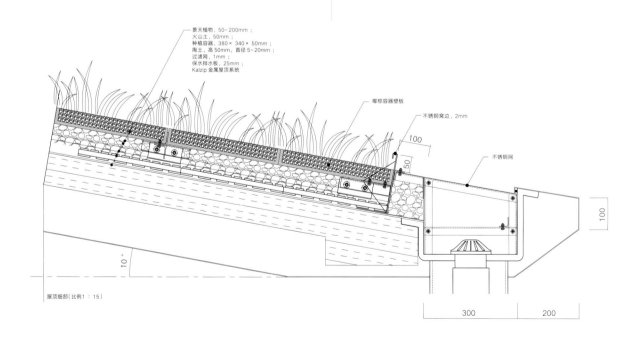

景天植物，50~200mm；
火山土，50mm；
种植容器，380×340×50mm；
陶土，高50mm，直径5~20mm；
过滤网，1mm；
保水排水板，25mm；
Kalzip 金属屋顶系统

椰棕容器壁板

不锈钢窗边，2mm

不锈钢网

屋顶细部（比例1：15）

祖先的智慧博物馆
Museum of Indigenous Knowledge

位置：菲律宾马尼拉市

主要用途：博物馆

总建筑面积：9,000m²（计划）

我们试图在马尼拉市中心做一个由岩石、土和绿色植物构成的巨大洞穴。从洞穴顶部出来能看到一片巨大的水面，水面上漂浮着传统的菲律宾民居。水、土和绿色植物是菲律宾文化的起源，这正是设计的主旨。

We are attempting to realize an enormous cave made from rocks, earth, and greens in the middle of the Manila metropolis. Upon emerging from the cave on the top level, one will encounter a cluster of traditional Philippine houses that will appear to float around an enormous water feature. In the design, I tried to convey the idea that earth, water, and greenery are the origins of Philippines culture.

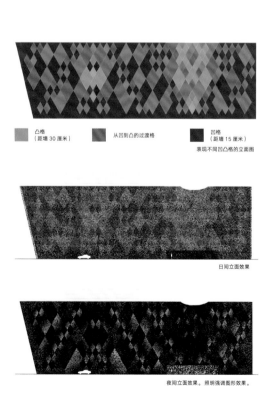

凸格
（距墙 30 厘米）　　从凹到凸的过渡格　　凹格
（距墙 15 厘米）

表现不同凹凸格的立面图

日间立面效果

夜间立面效果。照明强调图形效果。

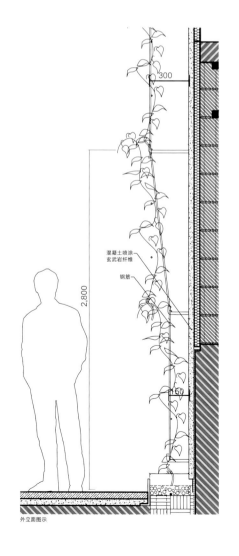

300

2,800

混凝土喷涂
玄武岩纤维

钢筋

外立面图示

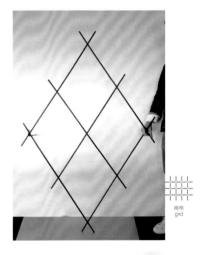

包装
wrapping

面向街道，我们想做一个覆盖爬藤植物的外立面。为此，我们用一种特殊的纤维去编织植被附着的基层框架，这种特殊材料是用熔化的玄武岩制成的，在菲律宾阳光的直射下也不会变热。

A special fiber made by melting basalt that does not get hot in the direct sunlight in the Philippines was woven in three dimensions to create a base for plants in order to create a facade covered with ivy that faces the street.

网格
grid

波特兰日本庭园文化中心
Portland Japanese Garden
Cultural Village

位置：美国俄勒冈州波特兰市

设计时间：2013.01—2015.07

建造时间：2015.08—2017.05

主要用途：文化中心、庭园、茶室、售票厅

总建筑面积：1,431.9m²

结构顾问：KPFF咨询工程公司

（KPFF Consulting Engineers）

建筑施工：霍夫曼建筑公司

（Hoffman Construction Company）

美国俄勒冈州波特兰市的日本庭园被认为是海外最好的日本庭园。在庭园入口的地方，我们在山坡上围绕广场做了这个文化中心。

为了让建筑与背后的山融为一体，我们用一种名叫greenbiz的多孔质、仅有30mm厚的轻量陶瓷板做屋顶，并做了绿化。greenbiz是一种用工业废弃物做的再生材料，原料是纤维染色过程中产生的废渣。我们设计了屋顶的坡度，使事先培育好的根部带土的景天植物能够直接固定在陶瓷板上。

Portland Japanese Garden, in Portland, Oregon, is considered the best Japanese garden outside Japan. At the garden's entrance, we created a center that surrounds a plaza and looks as though it were built into the hillside.

To integrate the buildings with the hillside, we greened the pitched roofs with greenbiz, porous ceramic panels, just 30 mm thick. (Greenbiz, made by drying out slag from a fiber dying process, are a kind of industrial waste.) We designed the roof with pitches that would allow sedum with soil on its roots to be affixed directly to the greenbiz.

包装
wrapping

我们研究了波特兰日本庭园中各种景天植物的生长状况，选择了最合适的植物组合。屋顶边缘的多孔陶瓷板只有30mm厚，看起来很轻盈。

The growth and development of various sedum in the Portland Japanese Garden were reviewed, leading to the optimum combination of sedum. The greenbiz panels on the edge of the roof are 30 mm thick, giving the green roof a light expression.

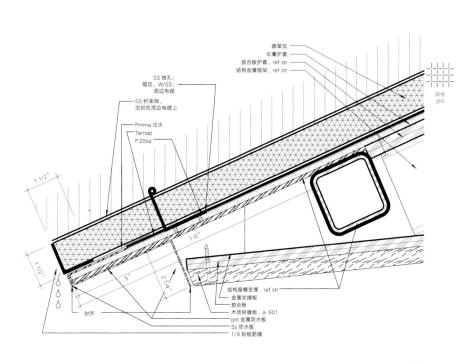

石

Stone

石头类似混凝土，又重又硬又冷，很长时间我都对它敬而远之。我特别不喜欢在混凝土建筑上覆盖薄薄的石材贴面——用材料贴面来给混凝土建筑做伪装，我不能忍受这种撒谎般的建筑手法。20世纪的建筑，总而言之就是在比拼混凝土上的材料贴面。

后来，我喜欢上石头的原因是安东尼奥·高迪设计建造的古埃尔领地教堂（始建于1908年，未竣工）。熔岩冷却后形成的六棱形岩柱，不经加工直接立在教堂的中心，作为支柱支撑屋顶。这样的石头简直就是树木。想想也是，石头原本就是大地的一部分，是土地的朋友，石头看上去和树木相似也没什么奇怪的。

之后我就开始了在设计中使用石头的各种挑战。就像古埃尔领地教堂那样，尽可能把石头作为一个块体来看待，而不是贴在混凝土表面的装饰素材。石头美术馆（2000）的业主是一家小型石材工厂的老板，因为没有预算，就只能用自己工厂加工的石材。业主希望建筑尽量不用混凝土、钢铁和玻璃，于是我们把石头作为一种固态的块体来看待，研究了很多种运用方法。

把石头重新定义为一种块状物，石头就不仅仅是表面有纹理的装饰材料，更是拥有各种内部组织的生物，像树木一样，生长着具有方向性的纤维，有正面和侧面的区分。

Stone resembles concrete. It is heavy and hard, and for a long time I avoided it. I especially disliked thin stone cladding over concrete. It was like decorating concrete with texture mapping, and I could not bear the feeling of dissimulation. In a word, the architecture of the 20th century put its stake on texture mapping over concrete.

What made me want to use stone despite my former dislike of the material was my encounter with Antoni Gaudí's Church of Colònia Güell (commenced in 1908, unfinished). There is a stone with a hexagonal shape called columnar basalt, which forms from slowly cooling lava. Unprocessed stone columns with that natural shape stood at the center of the church, supporting the roof. Although it was stone, it looked like it could have been a tree. When you stop to think about it, stone is also part of the earth. It is a friend of earth, so there is nothing strange about it being akin to trees.

Since then we have taken on the challenge of using stone in various new designs, but as far as possible as a block, like Church of Colònia Güell, and not as a surface cladding over concrete. In the Stone Museum (2000), the client was the owner of a quarry and wanted to use the limited budget on stone from the quarry instead of on concrete, steel and glass. With this brief, we were able to develop several ways to use stone as a block.

When stone is redefined as block, it starts to look like a living thing with various internal structures, instead of only as a material with a surface pattern. Like a tree, it has internal configurations, fibers with directionality, and straight and cross grains.

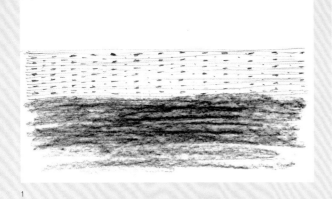

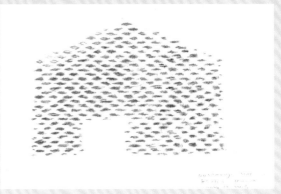

1

2

这样想，似乎生物与非生物之间的边界也消失了。树木不断纳入死去的躯体组织，变得牢固强硬，从这个意义上说，树木有一半是没有生命的。这个过程再进一步就变成了石头。所以，石头其实是有生命的。生与死的边界竟是如此模糊。建筑是徘徊在生与死的边界，能够让人意识到生死一体的一种存在。

From that viewpoint, the border between animate and inanimate things seems to vanish. Trees become harder and stronger by incorporating their own dead cells. In that sense, a tree is half dead. The same process carried a bit further yields a stone. Which means that stones are very much alive. The boundary between life and death is surprisingly vague. By wandering in the territory between life and death, architecture can show that life and death exist on a single continuum.

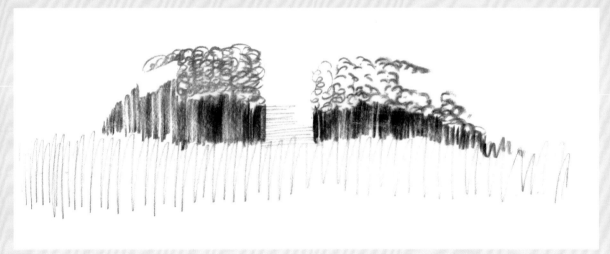

3

4

Sketches by Kengo Kuma
1. Stone Museum
2. Chokkura Plaza
3. Jeju Ball
4. V&A Dundee

隈研吾手稿
1. 石头美术馆
2. 巧珂垃广场
3. 济州球
4. 维多利亚和阿尔伯特博物馆·邓迪分馆

石头美术馆
Stone Museum

位置：日本栃木县

时间：1996—2000

主要用途：美术馆

占地面积：1,382.06m²

总建筑面积：527.57m²

合作建筑：茂木新一（Shinichi Mogi）

结构顾问：中田结构设计事务所

建筑施工：石原建设ECRIS

（Ishihara Construction ECRIS）

这是一个对大正时期的石米仓进行增改建的项目。原来的石米仓用的是一种名叫芦野石的凝灰岩，我们尝试用同样的石材做出轻盈、通透的建筑。

有些墙用截面40×120mm的棒状石材做成横向的格栅。还有些墙局部插入6mm厚的白色透光大理石，大约占整个墙面的1/3，实现了一种"透光的砖石结构"。芦野石在高温加热下会因为氧化还原反应而变色，设计茶室时，我们利用这种特性，在砖窑中烘烤石材，让石格栅的色彩和细节富于变化。

This project concerned the renovation and expansion of a stone rice storehouse built in the Taisho period. My goal was to create a light, transparent work of architecture using the same type of stone—a tuff called Ashino stone—used in the existing building.

On some walls, we formed stone louvers from stick-like pieces of stone, 40 × 120 mm in section. On other walls, we created a translucent masonry structure by stacking 6-mm-thick pieces of translucent white marble alongside the other stone, at a ratio of approximately one to three. When Ashino stone is baked at high temperatures, it changes colors due to a redox (reduction-oxidation) reaction. Exploiting this property, we baked stone in a brick kiln and used it in louvers and details, to create variety.

空　　　替代砖石　　　芦野石

石墙图样

粒子化
particlizing

堆叠
stacking

网格
grid

我们抽掉了 1/3 的石块，这样给砖石结构带来了透光性。在茶室的内部设计中，我们使用同样的芦野石，但通过不同的烧成温度得到了不同的色彩。

We removed one third of the stone blocks so that we could give transparency to the masonry. In the stone tea house, we applied the same Ashino stone that was changed to a variety of colors depending upon the firing temperature.

莲屋
Lotus House

位置：东日本
设计时间：2003.07—2004.04
建造时间：2004.06—2005.06
主要用途：别墅
占地面积：2,300.66m²
总建筑面积：530.27m²
结构顾问：橡树结构设计事务所
（Oak Structural Design Office）
建筑施工：松下工业（Matsushita Industry）

用销子将200×600×30mm规格的多孔大理石板材，固定在厚6mm、宽18mm的不锈钢条上，做成石屏，兼顾了透明感和石材的质感。石板呈棋盘状排列，营造了很好的通透感。

By attaching 30-mm-thick travertine panels (200 × 600 mm) to stainless steel flat bars (6 × 18 mm) using dowels, we created a stone screen that is both textured and transparent. We accentuated the impression of transparency by arranging the stone panels in a checkerboard pattern.

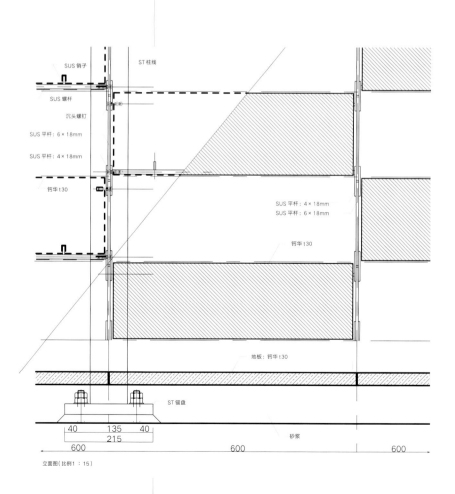

立面图（比例1：15）

粒子化
particlizing

编织
weaving

网格
grid

巧坷垃广场
Chokkura Plaza

位置：日本栃木县

设计时间：2004.03—2005.03

建造时间：2005.07—2006.03

主要用途：礼堂、展厅

占地面积：2,668.52m²

总建筑面积：728.18m²

结构顾问：橡树结构设计事务所

建筑施工：渡边总建（Watanabe General Construction）、见目石建（Kenmoku Stone Architect）

大谷石是一种多孔质、轻量的凝灰岩，为弗兰克·劳埃德·赖特所青睐，并用在了原东京帝国饭店（1923）的设计上。我们想用大谷石做成一种轻巧、多孔质的石屏。考虑到大谷石比较脆的结构特性，我们用6mm厚的钢板与大谷石编织在一起，做成一种混合结构的石屏。我们将这种石屏应用在原有的大谷石墙的改造上，让古老的仓库呈现出渐变透明的变化。

Oya stone is a type of porous pumice tuff that Frank Lloyd Wright loved and used in the Imperial Hotel (1923). Here, we used Oya stone to create light, porous stone screens. Mindful of the weak structural properties of the stone, we realized the screens using a mixed structure that diagonally interweaves stone blocks with 6-mm-thick steel plates. By creating gradients that transition from opaque to transparent, we used the screens to preserve a historic storehouse and to repair existing masonry walls made from Oya stone.

在结构顾问师新谷真人的指导下，我们就窗边的石屏做了实验模型，并做了负载强度测试。

A mock-up stone screen beside the window was made under the guidance of structural engineer Masato Araya, and a load was applied to conduct a strength test.

编织
weaving

支撑
reciprocating

堆叠
stacking

组装石材，然后架起钢板，接着再组装石材，这种交替
进行的施工方法是一个高难度的挑战。

We took on the challenge of a difficult building method
called "Aiban" in which stones are laid, steel plates are
placed on top, and more stones are laid.

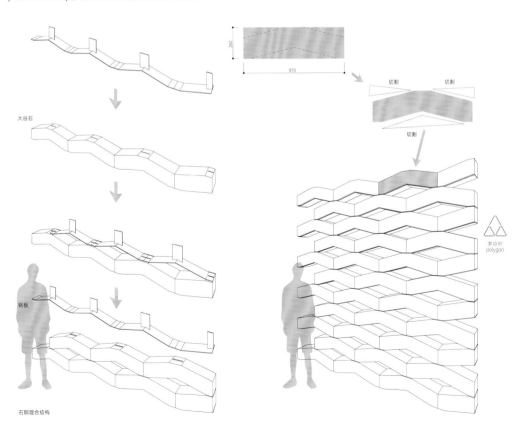

大谷石

钢板

石钢混合结构

多边形
polygon

石片城堡
Stone Card Castle

位置：意大利维罗纳市

时间：2007

主要用途：装置

总建筑面积：104m²

结构顾问：江尻建筑结构设计事务所

建筑施工：竹中工务店（欧洲）（Takenaka Europe）、坎诺比奥（Cannobio）

佛罗伦萨有一种布鲁内莱斯基和米开朗琪罗都很喜欢用的灰色砂岩——塞茵那石。我们用10mm厚的塞茵那石板搭建成一个可以是墙也可以是置物架的结构体，它如屏风般轻巧透明。这个设计的灵感来自用扑克牌搭建的"卡片城堡"。

同样的结构原理还用在了我们后来用铝板材做的"铝板拼接陈列架"项目中，成为一种更轻盈灵活的结构体系。"石片城堡"在意大利维罗纳市展出后，被拆成片，用卡车运到欧洲各地展出。

Borrowing a structural idea from "card castles" (houses of cards), we created an arrangement of light, transparent screens that function as both walls and furniture. The principle material is 10-mm-thick tiles of gray Pietra Serena sandstone, which was favored by Brunelleschi and Michelangelo.

In a subsequent project, Polygonium, we used the same structural principles but substituted the stone with extruded aluminum. This resulted in an even lighter and more flexible system. After being exhibited in Verona, the castle was disassembled into pieces, transported by truck, and displayed at various locations in Europe.

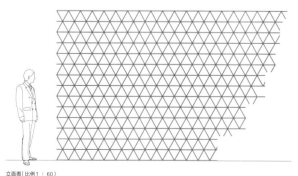

立面图（比例1：60）

细部（比例1：10）

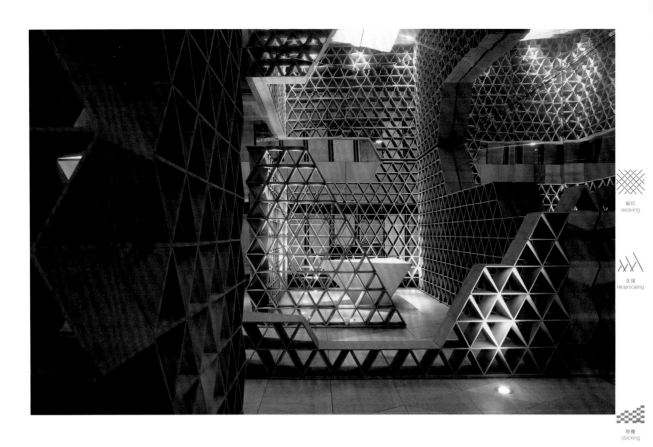

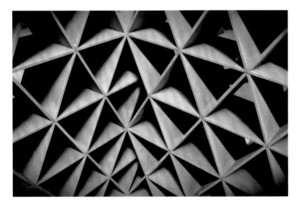

編織
weaving

支撑
reciprocating

堆叠
stacking

多边形
polygon

济州球
Jeju Ball

位置：韩国济州岛

设计时间：2009.02—2010.03

建造时间：2010.07—2012.03

主要用途：别墅、酒店

占地面积：13,181.98m²

总建筑面积：2,936.52m²

结构顾问：韩国未来结构顾问

（MIRAE structural engineers）

建筑施工：乐天工程建筑

（LOTTE Engineering & Construction）

韩国济州岛上有一种黑色圆球形的火山岩，我们用这种石块覆盖住整个屋顶，整片建筑群也围绕着"黑色的圆球石头"这个概念来设计。门窗上方的屋檐也是在不锈钢网板上直接铺石块。阳光像穿透密林那样穿过"石头屋檐"，洒下斑驳光影。

In this project, we covered all roof surfaces with round, black volcanic rocks from Korea's Jeju Island. This gave the buildings themselves a round, black appearance. The same black rocks were also stacked directly on top of stainless steel mesh above doors and windows. This produced "stone eaves" that transmit dappled sunlight like leafy trees in a forest.

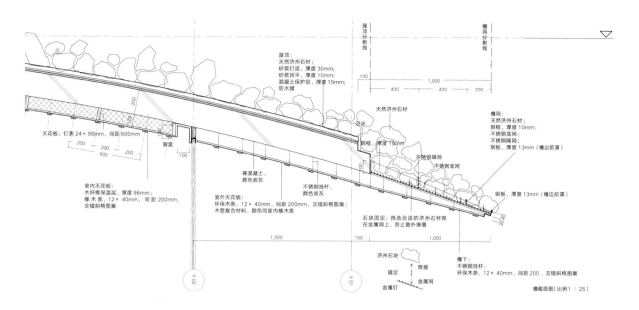

屋顶分割线　檐网分割线

屋顶：
天然济州石材；
砂浆打底，厚度30mm；
砂浆找平，厚度10mm；
混凝土保护层，厚度15mm；
防水膜

天然济州石材

泛水

檐网：
天然济州石材；
钢框，厚度10mm；
不锈钢底网；
不锈钢隔网；
钢板，厚度13mm（檐边前罩）

天花板：钉条24×90mm，间距600mm

盲盒

钢框，厚度10mm

不锈钢隔网

不锈钢底网

室内天花板：
木纤维保温层，厚度96mm；
橡木条，12×40mm，间距200mm，
交错斜格图案

室外天花板：
环保木条，12×40mm，间距200mm，交错斜格图案；
木塑复合材料，颜色同室内橡木条

裸混凝土，
颜色炭灰

不锈钢挂杆

钢板，厚度13mm（檐边前罩）

石块固定：挑选合适的济州石材焊
在金属网上，防止意外滑落

檐下：
不锈钢挂杆
环保木条，12×40mm，间距200，交错斜格图案

济州石块
锚定　　焊接
金属钉　金属网

墙截面图（比例1：25）

在济州岛，农民用火山岩砌农田的防风墙，整个岛都被
赋予了火山岩的质感。我们把这种独特的质感延续到了
建筑上。

The volcanic rock on Jeju Island is used for windbreak
walls in fields, and the entire island is covered with this
volcanic rock texture. This unique texture was extended
into architecture.

Stone 石   **149**

维多利亚和阿尔伯特
博物馆·邓迪分馆
V&A Dundee

位置：英国苏格兰

时间：2010—2018

主要用途：美术馆、教育设施

总建筑面积：8,500m²

合作建筑：CRE8建筑事务所（CRE8
ARCHITECTURE）、彼得·鲍曼（Peter
Bowman）、林恩·阿尔加（Lynn Algar）

结构顾问：奥雅纳工程顾问公司（ARUP）

苏格兰北部，泰河流过邓迪市，维多利亚和阿尔伯
特博物的邓迪分馆就在河口临水而建。建筑像悬崖
一样向水面伸展出去，这是苏格兰海边的粗犷的条
纹状悬崖给我们带来的灵感。预制混凝土的条块之
间留有间隙，使外立面获得一种阴影效果。

我们计算了光线的效果，设计出截面呈梯形的混
凝土构件。粗糙的混凝土骨料，再加上高压喷
砂——我们尽可能在建筑这种人造物中把自然的
悬崖所具有的力量感和粗糙感表现出来。

This museum in Dundee, northern Scotland, is situated
at the mouth of the River Tay and extends out over
the water. Drawing inspiration from the rough, striated
cliffs in Scotland that divide the land from the sea,
we created a shaded facade composed of precast
concrete members spaced out to let light in.

The precast concrete members are trapezoidal in
section based on calculations of lighting effects and
were made with a rough aggregate that has been
revealed through pressure washing. My aim was to
imbue this man-made work of architecture with the
strength and roughness of the natural cliffs.

（上）三维曲面的混凝土构件模具先在德国用钢板制作，再用船运至邓迪。

（下）用粗糙的碎石做混凝土骨料，再通过喷砂最大限度地使骨料暴露出来，以呈现苏格兰悬崖的粗犷质感。

(Above) The molds for pouring concrete with three-dimensional curves were made in Germany using steel plates, and transported to Dundee by ship.
(Below) Precast concrete was made using rough crushed stone as the aggregate, and sandblasted to expose the aggregate to the maximum extent possible, achieving a rough texture that evokes an image of the cliffs in Scotland.

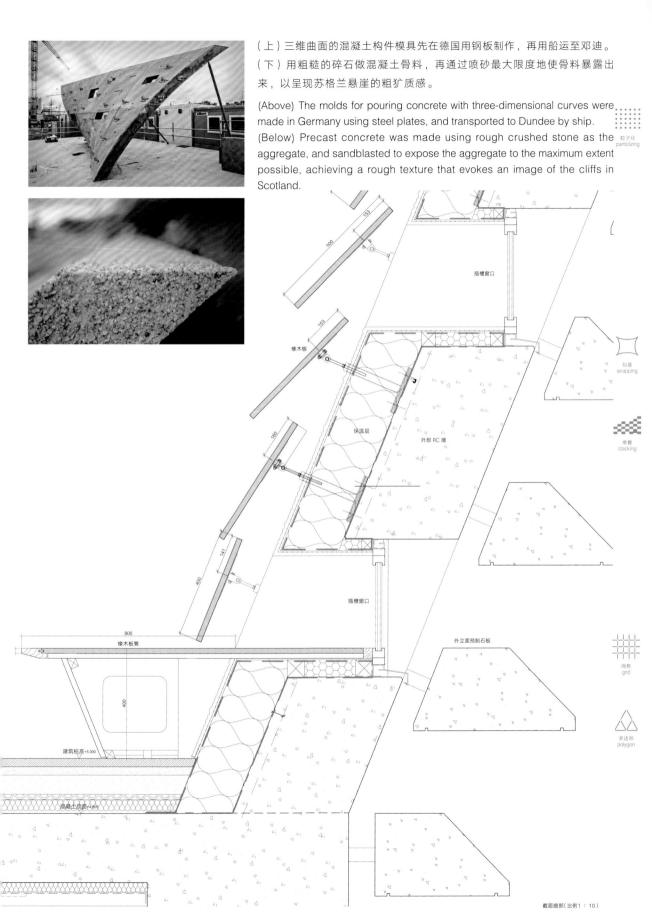

粒子化
particlizing

包装
wrapping

堆叠
stacking

网格
grid

多边形
polygon

插槽窗口

橡木板

保温层

外部 RC 墙

插槽窗口

外立面预制石板

800

橡木板凳

400

建筑标高 +5.000

混凝土面标高 +4.900

153
500
153
160
141
450

截面细部（比例 1：10）

火：

陶瓷砖瓦／玻璃／树脂

Fire: Tile / Glass / Resin

我把经过火的作用从液态变为固态的材料归入一组。火是家的中心，不仅提供温暖，保护人的生命，还能作用于各种物质，让物质发生改变。颗粒状的土在火的力量下变为液态，进而有了转化的可能。火有着魔术般的力量，因为火，物质在固态、液态、气态之间来回转化。戈特弗里德·森佩尔把建筑的工作分为三类：大地、火、编织。

提出"火"的概念，在当时是很出人意料的，因为关于火，一般人只会想到"炉火"，它只是建筑中的一个配角。可是，如果把建筑行为看作一种物质转化的过程，那么火就不仅仅意味着炉火，更应该是建筑行为的中心。我们人类的身体中流淌着温暖的液体，对火和液体总有某种亲近感。玻璃不仅仅意味着一个透明的平面，它还是土，也是火和液体。树脂也是这样，如果从火和液体的角度去定义树脂，那么我们对树脂也会感到很亲切。

布鲁诺·陶特就是因为认识到玻璃是土、火和液体，才有了与水结合的玻璃馆（1914）的设计。我也把水与玻璃相结合，在陶特设计的日向邸的旁边设计了"水／玻璃"（1995）。陶特的玻璃馆之后，密斯·凡·德·罗做了只有玻璃的透明建筑群。人们忘记了玻璃是土，是火，也是液体。我觉得，是时候重新把玻璃当作一种有生命的东西来对

This issue defines a group of materials that begin as liquids and are transformed to solids by the application of fire. Fire has a central place in the house. Not only does it provide life-protecting warmth, it also operates on various materials, transforming them into something different. Fire has the power to transform earth particles into liquid and different kinds of material. Thanks to this magical power, materials can go back and forth between their liquid, solid, and gas states. Gottfried Semper divided the work of architecture into three categories, related to earth, fire, and weaving.

At first it seemed strange to find fire in this list, since the hearth seemed to play only a supporting role in architecture. But if architecture is viewed as the story of material transformations, then both the hearth and fire itself deserve to be positioned at the center of the architectural endeavor. We probably feel an affinity for fire and liquids because our bodies themselves are streams of warm liquid. Glass is not only a transparent surface. It is also earth, fire, and liquid. Resin can also be viewed as fire and liquid, which makes it feel closer.

Bruno Taut knew that glass is earth, fire, and liquid, which is why he combined water and glass in his Glass Pavilion (1914). I have also combined water and glass in my Water / Glass project (1995), which stands next to the Hyuga Villa (1936) designed by Taut. After Taut's Glass Pavilion, Ludwing Mies Van der Rohe designed transparent pavilions of glass only. The idea that glass is earth, fire, and liquid was forgotten. But I feel that the time has come to recover glass as a living thing.

待了。

　　树脂也是这样，回到它的起源去看待它，就会感到亲切。就像"树脂"这个名称所揭示的，它原本是堆积在海底的植物、藻类等，后来变成了石油，然后我们用石油做成了树脂。想到树脂的来历，我们便有理由对它产生亲近感。因为有着这样的亲切感，我们挑战了树脂各种各样的成形方法，也尝试用树脂结构去组建建筑。

When viewed in the same way—thinking back to the origins—resin also seems closer. As indicated by the name, resin was originally plant life and seaweed that sank to the bottom of the sea and became petroleum, from which it is made. From this history, it is not without reason that we feel close to resin skins. To take advantage of that feeling of closeness, we are working on the challenge of incorporating it into architecture, using materials made with resin cells.

1

2

3

4

Sketches by Kengo Kuma
1. Xinjin Zhi Museum
2. Tiffany Ginza
3. FRAC Marseille
4. Oribe Tea House

隈研吾手稿
1. 新津·知美术馆
2. 蒂芙尼银座
3. 马赛当代艺术中心
4. 织部茶室

分德山
Waketokuyama

位置：日本东京
设计时间：2003.01—2003.08
建造时间：2003.09—2004.02
主要用途：餐厅
占地面积：245.67m²
总建筑面积：149.66m²
结构顾问：橡树结构设计事务所
建筑施工：Seijimo建设工业
（Seijimo Kensetsu Kogyo）

对水泥、硅酸原料和纤维原料进行热处理以提高硬度，然后挤压成水泥板，再将水泥板切割成条状。为了减轻水泥板的重量，在其内部留有孔洞。用水泥板条装饰外墙时，特意将孔洞暴露出来。这些孔洞除了能给外墙带来透明度，其产生的阴影还改善了工业材料冷硬的质感，使之变得柔和温暖，更符合日本料理餐厅的气氛。

Cement, silicic acid raw materials and fibrous raw materials were heat treated to increase their hardness and extrusion molded cement boards were sliced. The holes in these slices that were made to minimize weight are visible when they are used as exterior walls. In addition to providing the exterior walls with transparency, the holes create shadows, transforming an industrial material traditionally considered to be cold and hard into something that is soft and warm, making it suitable for a Japanese restaurant.

颗粒化
particlizing

堆叠
stacking

网格
grid

吴市音户町市民中心
Kure City Ondo Civic Center

位置：日本广岛市
设计时间：2004.07—2006.02
建造时间：2006.06—2007.12
主要用途：市政厅分局、市民活动中心、多功能厅
占地面积：4,424m²
总建筑面积：4,642m²
结构顾问：橡树结构设计事务所
建筑施工：日光公司和鸿池运输联合
（JV of Konoike and Nikko Corporation）

濑户内海的这个小岛上，有着成片的瓦屋顶住宅。我们希望新建的公共建筑能够融入这种具有怀旧感的风景中，因此在使用传统"本瓦"（凹凸两种瓦的组合）的基础下，做了一些细节的创新。

江户时代才开始流行的"栈瓦屋顶"（仅用一种波浪形的瓦铺砌）看上去比较平，形态比较单一。而用凹凸两种瓦组合铺砌的"本瓦屋顶"有更深的阴影，还能体现出手工制作的丰富质感。在音户一带，这种本瓦屋顶也是城镇景观的底色。我们将本瓦中的凸瓦与钢架结合，得到了一种能够透光的百叶窗状的屋顶，创造了一个光线舒适的公共广场。

We took on the challenge of creating new details in traditional tile with the objective of making a public building blend into the nostalgic scenery of small islands in the Seto Inland Sea where there are many homes that have tile roofs.

Compared to the flat look of uniform pantile that spread during the Edo era, traditional tile roofs that combined two types tile that are convex and concave have deep shadows and a rich handmade texture, making the Ondo Civic Center a key part of the scenery in the Seto straits. Furthermore, combining the convex portion of the traditional tiles with steel beams enabled the creation of a louver shaped roof that allows light to pass through, creating a public square where light is controlled that is fitting for the Seto area.

颗粒化
particlizing

围合
wrapping

网格
grid

陶瓷之云
Casalgrande Ceramic Cloud

位置：意大利雷焦艾米利亚
时间：2009.01—2010.09
主要用途：纪念性装置
占地面积：2,697m²
结构顾问：江尻建筑结构设计事务所

我们不在混凝土上贴瓷砖，而是用瓷砖本身做结构体，设计了一个透明的陶瓷纪念性装置。横向上，600×1,200×14mm的陶瓷板夹着玻璃纤维网片延伸；纵向上由直径20mm的不锈钢管支撑。陶瓷板与不锈钢管编织成一个高5.4m、长45m的结构体，融入意大利的草原。其两端只有一片板的厚度，线条锐利，从远处看仿佛只有一根细线。

A transparent tile monument was built by using the tiles as the structure rather than attaching the tiles to concrete. Weft members consisting of 14 mm thick ceramic tiles measuring 600 × 1,200 mm that sandwich glass fiber mesh were woven together with woof members made from stainless steel pipe with a diameter of 20mm to create a 45 meter long monument that is 5.4 meters high so that it blends in with the surrounding grass fields in Italy. Both ends have pointed tiles that makes the monument appear to be a single thin line when viewed from a long distance.

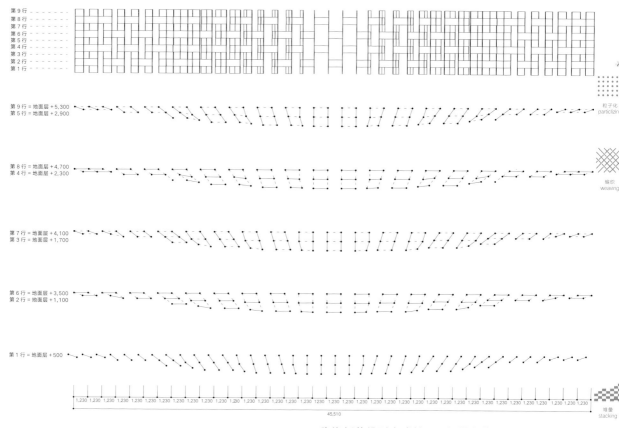

第9行 ------
第8行 ------
第7行 ------
第6行 ------
第5行 ------
第4行 ------
第3行 ------
第2行 ------
第1行 ------

第9行 = 地面层 +5,300
第5行 = 地面层 +2,900

第8行 = 地面层 +4,700
第4行 = 地面层 +2,300

第7行 = 地面层 +4,100
第3行 = 地面层 +1,700

第6行 = 地面层 +3,500
第2行 = 地面层 +1,100

第1行 = 地面层 +500

1,230 ×... 45,510

粒子化
particlizing

编织
weaving

堆叠
stacking

网格
grid

多边形
polygon

陶瓷板的排列方式基于几何学上的考虑，观看的视角不同，透明度也会不一样。

The geometry of the ceramic panels was determined in such a way that transparency changes depending upon differences in the point of view.

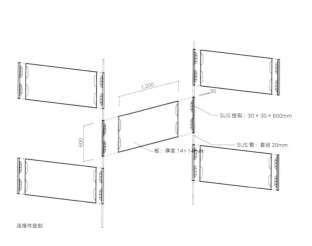

1,200
30
600
SUS 挂钩：30 × 30 × 600mm
SUS 管：直径 20mm
板：厚度 14+14mm

连接件细部

新津·知美术馆
Xinjin Zhi Museum

位置：中国成都
时间：2008—2011
主要用途：艺术馆
占地面积：2,580m²
总建筑面积：2,353m²
结构顾问：橡木结构设计事务所
建筑施工：中国二十冶集团有限公司（China MCC20 Group）（建筑）、深装总建设集团（Shenzhen Decoration & Construction: Industrial ）（幕墙）

道教圣地老君山脚下的这座艺术馆，是以瓦为主题进行设计的。本地烧制的瓦片质地粗糙，我们在瓦片的四角开孔，用铁丝将瓦片固定在不锈钢索上。利用双曲抛物面扭壳的几何学原理，用直线的不锈钢索做出了柔和的曲面。挡墙部位的表面处理也用同样的瓦片。整个建筑都统一在瓦的质感中。

Tile was used for a museum at the base of Laojun Mountain, a holy place for Taoism. Holes were made in the four corners of tile made with a local field burning method that have a rough random look, and the tile was secured to stainless steel wire with annealing wire. The geometry of hyperbolic paraboloid shells that generate curves with straight lines was used to create soft curves while using straight stainless steel wire. Tile with the same finish was attached to the retaining walls, covering the entire building with the texture of this tile.

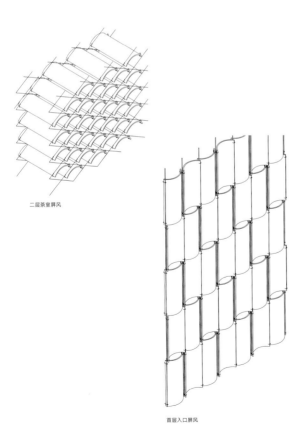

二层茶室屏风

首层入口屏风

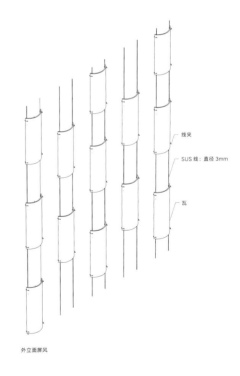

线夹

SUS 线：直径 3mm

瓦

外立面屏风

在菱形编织的钢网中插入本地瓦片，做成的屏风让人想到传统的青海波图案。

A stainless steel wire mesh woven in a diagonal pattern was filled with locally made roof tile in order to achieve a screen that evokes an image of a traditional pattern Seigaiha.

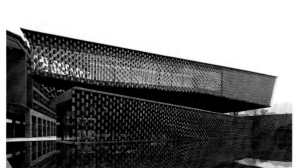

粒子化
particlizing

编织
weaving

包装
wrapping

网格
grid

螺旋
spiral

中国美术学院民艺博物馆
China Academy of Art's Folk Art Museum

位置：中国杭州
时间：2009—2015
主要用途：艺术博物馆、会议室
占地面积：9,525m²
总建筑面积：4,556m²
结构顾问：小西泰孝建筑构造设计
（Konishi Structural Engineers）

博物馆所在的场地是一个缓坡，原本是一座茶园。沿着坡地，我们设计了一座向斜下方延伸的单层建筑。我们用多边形分割的方法，将地形复杂的斜坡分割成很多三角形。多个三角形的小屋顶聚集在一起，看上去就像一个村落。

屋顶与遮挡阳光的屏风般外墙用的是本地烧的瓦。瓦的粗糙质感和色差正是我们想要的效果。新津的知美术馆是把瓦片固定在平行排布的钢索上，在这里，钢索被编织成菱形的格子，瓦片的分布可以更加随机自然。瓦片的尺寸是230×300mm，从远处看也很大，这考虑了粒子与整个建筑规模之间的平衡。

A one-story museum with continuous sloping floors was built to hug the gentle slopes of a site that was a tea field in the past. Polygon divisions were used to divide the museum into a collection of complicated inclined triangular shapes to give the small triangular roofs of the museum the appearance of houses in a village.

Tiles made locally with a field burning method were used for the roofs and exterior wall screens to block out direct sunlight, and the roughness and variations between the tiles were effectively utilized. The local field burnt tiles were secured to stainless steel wire strung in a parallel pattern in the same manner as at the Xinjin Zhi Museum, but here stainless steel wire was also woven in a diagonal pattern to enable the tiles to be arranged in a more random pattern. The tiles measure 230 × 300 mm, making them appear large even from quite a distance, creating a feeling of particles that is balanced with the scale of the building.

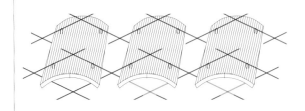

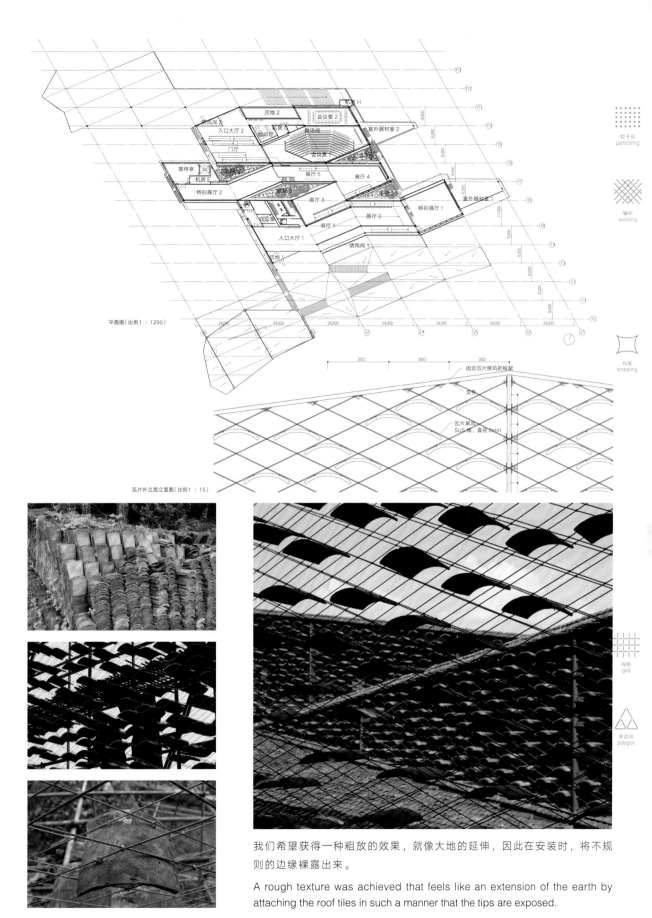

平面图（比例1：1200）

瓦片外立面立面图（比例1：15）

300   300   300

固定瓦片屏风的框架

瓦片屏风
SUS 线，直径 2mm

粒子化
particlizing

编织
weaving

包裹
wrapping

网格
grid

多边形
polygon

我们希望获得一种粗放的效果，就像大地的延伸，因此在安装时，将不规则的边缘裸露出来。

A rough texture was achieved that feels like an extension of the earth by attaching the roof tiles in such a manner that the tips are exposed.

船厂 1862
Shipyard 1862

位置：中国上海
设计时间：2011.12—2014.04
建造时间：2014.05—2017.05
主要用途：零售店、剧院、多功能厅
占地面积：11,340m²
总建筑面积：31,626m²
合作建筑：上海建筑设计研究院（SIADR）
结构顾问：奥雅纳工程顾问公司（上海）
建筑施工：上海建工集团（Shanghai Construction Group）

这里原本是上海黄浦江边的一座砖砌造船厂，我们尽可能地保留了它巨大的厂房空间，特别是那些独特的预制混凝土桁架结构，让它们作为空间的主角暴露出来。外墙的砖色彩不均衡，有明显的手工感，也被我们保留下来。新做的墙则设计了一种特殊的细节，把不同色调的中空陶砖固定在不锈钢拉杆上，用新的技术手段去模拟那种手工感。陶砖分布的密度也有变化，越靠近江边密度越大。

A shipyard built with bricks along the Huangpu River in Shanghai was restored in order to preserve the huge space in the facility as much as possible, and feature the unique pre-stressed concrete truss structure so that it is visible. The bricks used for the exterior walls that have irregular colors and a handmade touch were preserved, and a special technique was used in which hollow terra cotta tiles with different colors were secured to the newly built wall in an attempt to replicate the handmade touch with new technology. The density of the terra cotta tiles was changed in a gradational manner, with the density increased the closer the tiles were to the river.

弯曲的光纤，给工业化的空间带来柔软。

Optical fiber was twisted in order to provide the industrial space with softness.

原有建筑的粗犷风格来自色调不一的砖，我们用不同色调的陶砖再现了这种粗犷。

The roughness of the existing architecture that was created by the bricks that have varying colors was recreated with terra-cotta tiles that have varying colors.

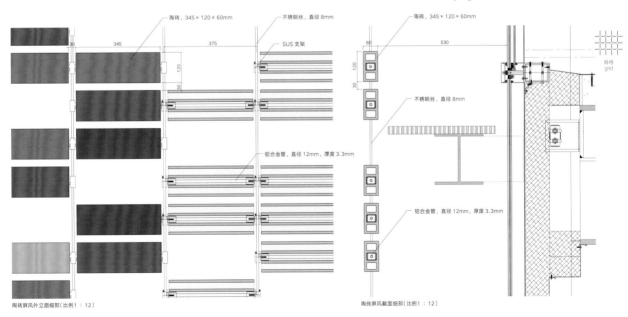

陶砖，345×120×60mm   不锈钢丝，直径8mm

SUS支架

铝合金管，直径12mm，厚度3.3mm

陶砖，345×120×60mm

网格
grid

不锈钢丝，直径8mm

铝合金管，直径12mm，厚度3.3mm

陶砖屏风外立面细部（比例1：12）

陶砖屏风截面细部（比例1：12）

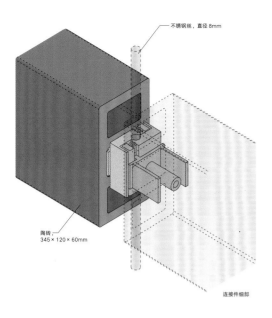

不锈钢丝，直径 8mm

陶砖：
345×120×60mm

连接件细部

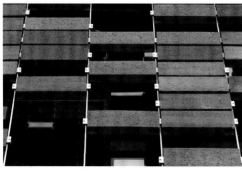

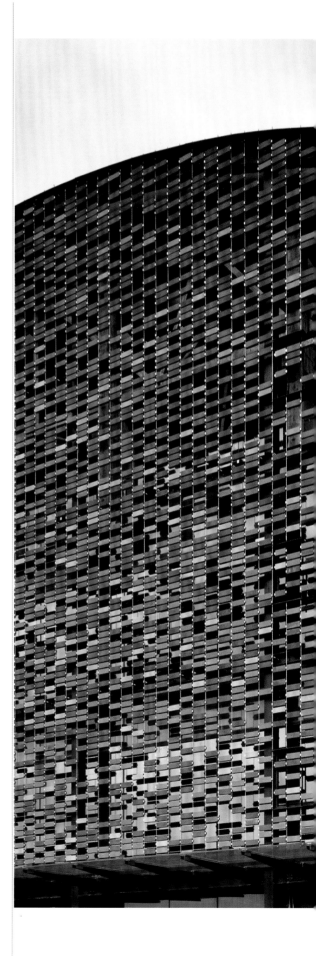

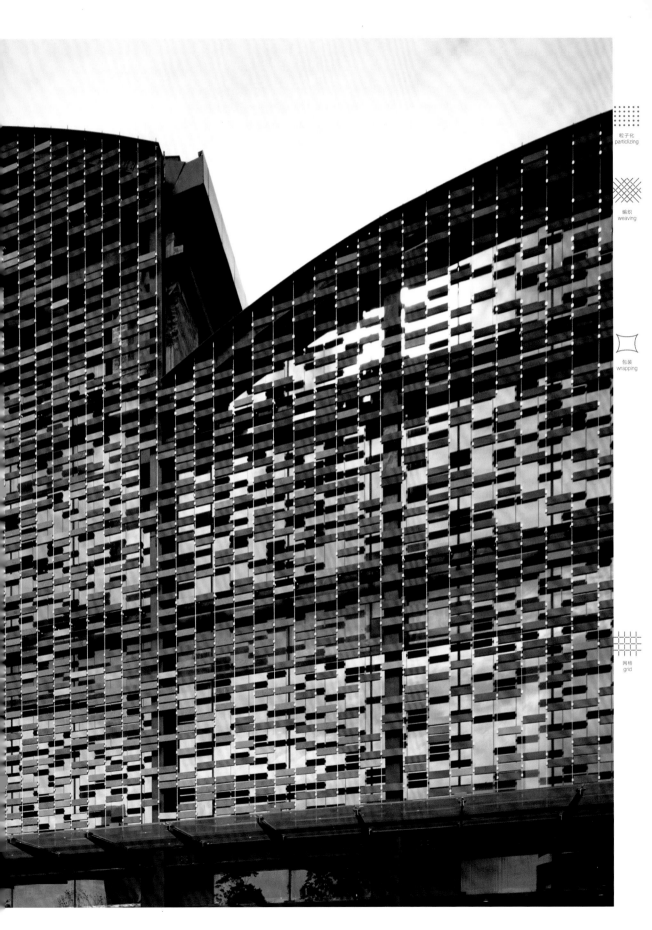

粒子化
particlizing

编织
weaving

包装
wrapping

网格
grid

蒂芙尼银座
Tiffany Ginza

位置：日本东京

设计时间：2007.08—2008.03

建造时间：2008.03—2008.10

主要用途：零售店、写字楼

占地面积：702.8m²

总建筑面积：5,870m²

结构顾问：橡树结构设计事务所

建筑施工：大成建设、三越环境设计（Mitsukoshi Kankyo Design）

把用于飞机机身的轻量、强度高的铝蜂窝芯材（厚10mm）夹在玻璃中间，做成外墙板，既通透，又能有效遮挡阳光。用汽车掀背上的铰链把墙板按不同的角度安装在原有建筑的外墙上，使整个建筑外立面闪烁着形态各异的光点。

Sustainable exterior wall units were developed that are made by sandwiching aluminum honeycomb core (thickness: 10 mm) used for aircraft bodies that is lightweight and strong between sheets of glass. These wall units feature high permeability and are also highly effective in shutting out solar radiation. Hinges used on the hatchbacks of cars were utilized to mount these wall units to the exterior wall of the existing building at differing angles, transforming the overall facade into a collection of random shining particles.

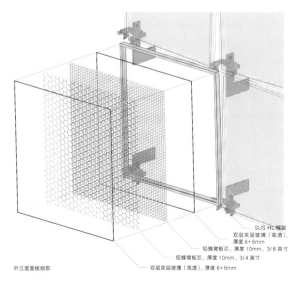

SUS HL框架
双层夹层玻璃（高透），厚度6+6mm
铝蜂窝板芯，厚度10mm，3/8 英寸
铝蜂窝板芯，厚度 10mm，3/4 英寸
双层夹层玻璃（高透），厚度6+6mm

外立面面板细部

使用铝蜂窝芯材，面板的通透性从中心向四周逐渐降低。

The use of aluminum honeycomb transforms the permeability of the center portion gradationally into an opaque surface.

粒子化
particlizing

包装
wrapping

网格
grid

切割铝质蜂窝芯材时用的是一种特殊的手推刀具，由熟练的技工手工操作。

Cutting of the aluminum honeycomb was performed by skilled craftsmen using a blade called a "hand push cutter."

马赛当代艺术中心
FRAC Marseille

位置：法国马赛
时间：2007—2013
主要用途：艺术中心
占地面积：1,570m²
总建筑面积：5,757m²
合作建筑：图里和瓦莱事务所（Toury et Vallet）
结构顾问：CEBAT工程（CEBAT ingénierie）

涂油处理的玻璃中间夹膜，并将其用作外墙的材料，从而获得日本传统纸窗那样半透明的效果。玻璃通过灵活的铰链安装在建筑外墙上，可以根据阳光调整角度，抵御法国南部强烈的日照；它还可以调节外部及内部观者的视线，成为舒适的不规则外立面。这样的外立面，使建筑具备复合性和灵活性的特点，可以满足当地年轻艺术家的多种需求，比如在其中生活和创作。

EVASAFE sheets (interlayer) were laminated between glass sheets with an enamel coating in order to develop exterior wall units that have a translucent effect similar to Shoji screens. These units were attached to the exterior walls with flexible hinges to allow adjustment to the desired angle as a means to control the strong sunlight in Southern France as an integral part of a sustainable random facade design that controls the line of sight from both the outside and inside of the building. This created a facade system that enables the composite and variable qualities of the building to be freely changed for local young artists to include apartment and studio configurations.

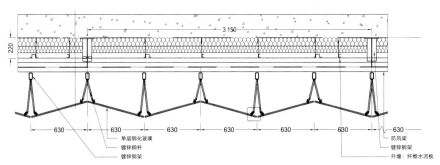

混凝土结构上的屏风水平截面图（比例1：40）

单层钢化玻璃
镀锌钢杆
镀锌钢架
防风梁
镀锌钢架
外墙：纤维水泥板

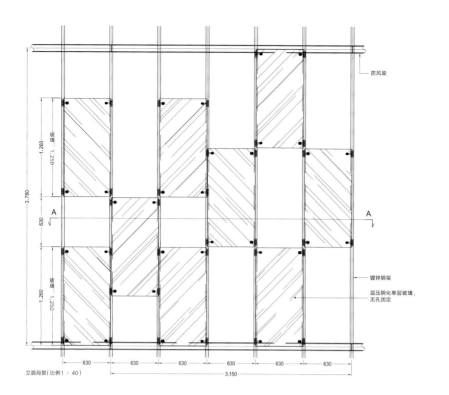

玻璃
1,250

A

630

玻璃
1,250

3,780

防风梁

镀锌钢架

层压钢化单层玻璃，
无孔固定

630 630 630 630 630 630

3,150

立面局部（比例1：40）

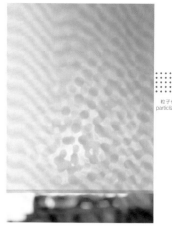

粒子化
particlizing

釉料涂层尽可能做得不均衡，使
整个外立面看起来自然不刻板。

The enamel coating process
was performed in a random
manner to the maximum extent
possible, achieving randomness
in the overall facade.

包装
wrapping

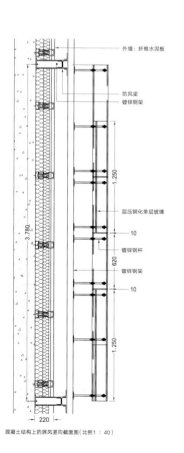

外墙：纤维水泥板

防风梁
镀锌钢架

层压钢化单层玻璃

3,780

1,250

620

镀锌钢杆

镀锌钢架

10

10

1,250

220

混凝土结构上的屏风竖向截面图（比例1：40）

网格
grid

火烧木系列
Yakisugi Collection

位置：任意
设计时间：2016.12—2017.01
建造时间（生产时间）：2017.02—2017.09
主要用途：吊灯
制造商：LASVIT灯具（LASVIT）

捷克是世界玻璃生产中心之一。我们与捷克的玻璃工厂合作，制作了这一系列富有立体感的玻璃产品。

直接将火烧木的纹理印到玻璃上，每一件产品都拥有独特的纹理。在当代建筑中，玻璃只是一种没有个性的工业产品，我们想用独一无二的手工玻璃产品改变这种状况。

This is a glass product with a three-dimensional texture that was made in collaboration with a glass factory in the Czech Republic, one of the centers of the glass industry in the world.

We developed a method to use pieces of Yakisugi (charred timber) as female dies and directly transfer the pattern to glass so that all products have a different texture, taking on the challenge of producing one-off handmade glass products, a material that is normally considered to be a uniform product in contemporary architecture.

每一件产品的模具，都是捷克小村庄诺维堡的工匠们用不同纹理的火烧木制作的。

Each individual piece is made with a different texture of charred cedar mold by craftsmen in Nový Bor, a small village in the Czech Republic.

织部茶室
Oribe Tea House

位置：日本岐阜县
时间：2005.02—2005.04
主要用途：茶室
总建筑面积：8m²

这是一个可移动的半透明茶室。5mm厚的塑料板切割成不规则的半圆形，间隔65mm平行排列，板与板之间用绑带固定。以"织部"命名茶室是对茶人古田织部的致敬，设计灵感来自他著名的变形茶碗。我们后来在罗马、北京和巴黎也做过这个茶室，材料改用当地的PC板（聚碳酸酯板，一种常用的塑料建材）。

A movable translucent tea house was built by using 5 mm thick plastic cardboard pieces that were cut in a distorted semicircular shape, lining them up parallel to each other through a space measuring 65 mm, and using tie bands to secure the pieces to each other. The tea house was named as an expression of respect for the distorted tea bowls for which Oribe Furuta is known. Tea houses were replicated in Rome, Beijing and Paris, with the material changed to polycarbonate plates which were locally obtained.

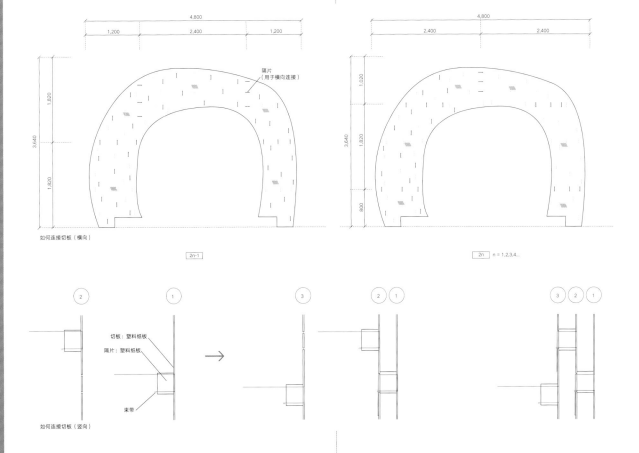

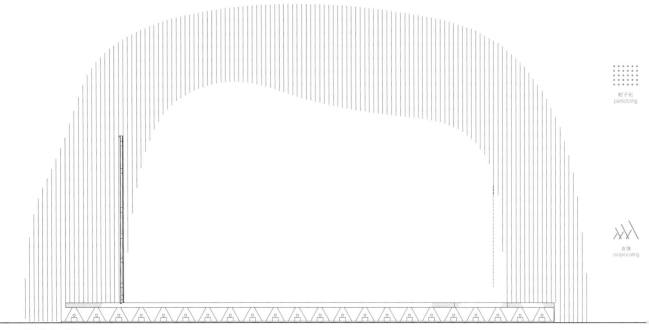

截面图（比例1：150）

粒子化
particlizing

支撑
reciprocating

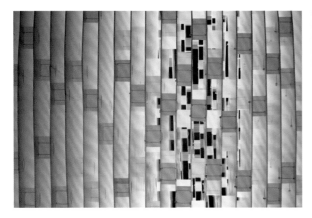

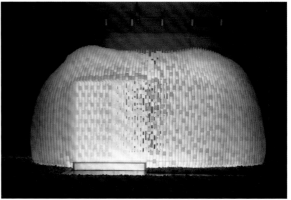

网格
grid

依据 CAD 软件输出的图形，手工切割塑料板。

Cutting of the corrugated plastic boards was all performed by hand using molds output with CAD software.

螺旋
spiral

水砖屋
Water Branch House

位置：美国纽约
时间：2008
主要用途：装置

"水砖"系列的灵感来自施工现场的临时路障——一种注水的塑料容器。最初我们做了乐高积木形状的水砖（2007年，东京六本木"间"美术馆，本页照片），开发了一种由水控制结构重量的外墙系统。在纽约现代艺术博物馆举办的"送货上门"展（2008年，191页照片）中，水砖被做成树枝一样的细长形状，并组成悬臂结构；水砖相互连接，可以灌入水。

在水砖组成的墙、地面和屋顶中注入温水或冷水，可以让建筑本身成为一种调节环境的装置。在现代建筑体系中，结构、内装和设备是垂直划分的独立领域，我们则试图找到一种融合统一的方式。如果在外部装上太阳能集热设备，水砖屋将具有很强的抗灾性能，不依赖外部基础设施，成为一种自给自足的系统。

堆叠
stacking

After obtaining a hint from the plastic containers into which water is poured and used as temporary barricades at worksites, an exterior wall system was developed for which the weight is changed with water, starting with Lego block shaped plastic containers (2007, exhibition at TOTO GALLERY · MA). In the Water Branch House that was exhibited at the MOMA Home Delivery Exhibition (2008), the units were made long and thin like branches, enabling a cantilever structure, the units were connected to each other, and water was poured inside.

网格
grid

The building structure can function as a type of environmental device by circulating warm water or cold water inside the walls, floor and roof. The objective of this design was to achieve a new type of integration that could replace modern building systems in which the structure, interior decorations and equipment are vertically divided. A self-sustaining system that is highly resistant to disasters and not dependent on external infrastructure can be created if a solar heat collection system is installed on the outside.

水砖是用塑料做的水箱，非常轻，易搬运。将其堆叠起来，可以自由搭建房子或家具。水砖里可以注入水或其他液体。一块水砖由5个100×100×100的方块错位组成，多块水砖可以有各种各样的组合方式。水砖凹凸咬合，还可以成为牢固的结构体。
Water branch is a piece of plastic tank. By piling them up, you can build anything from furniture to a house. It is very light and easy to carry around. Water or other types of liquid can be stored inside. It is in the shape that each cube of 100 × 100 × 100 mm is connected staggeringly so they can be turned into a variety of shapes. Furthermore, it can form a strong structure by joining its concave and convex firmly.

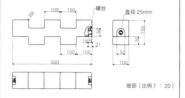

细部（比例 1：20）

接合结构
Joint system

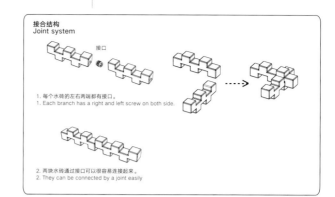

接口

1. 每个水砖的左右两端都有接口。
1. Each branch has a right and left screw on both side.

2. 两块水砖通过接口可以很容易连接起来。
2. They can be connected by a joint easily

生态材料
Ecological material

水·二酸化炭素

我们尝试用一种生物降解或水解的聚酯做水砖，如果成功，就能实现一个从容器、建筑材料到土的可持续循环系统。
Water branch is a trial case of using PET, the Hydro/Biodegradable polyester that can eventually go back to the ground. If it is successful, a new sustainable recycling system will be realized that takes the route from a container, to construction material, and to soil.

循环系统
Circulation system

1. 在水砖中注入水或其他液体，可以调节重量。
1. The weight of Water Branch can be adjusted by the volume of liquid that you pour inside.

2. 水砖相互连接，液体可以在内部自由流动。
2. By connecting the pieces, liquid can flow into the next branch and run around within the tanks.

3. 因此，水砖系统不仅是一种结构体，还具备多种功能：
3 Water Branch can function not only as a structure but also as many other roles:

注入温水或冷水成为控温系统；
网络排线；
利用水箱的凹凸结构，做成过滤净水系统；
柔软材料能吸收震动；
照明设备；
收储雨水；
墙面和地面绿化；
注入不同的材料（土、沙、混凝土、不透明液体等），得到不同的特性。

thermal insulation,
network wiring,
filtering by concave and conves, water purification system with precipitation tank,
absorbing shock with its soft material,
lighting equipment,
soring rainwater,
greening of wall and floor,
change its role by the thing you put inside (such as mud, sand, concrete, opaque liquid, etc.).

装配形式
Assembling pattern

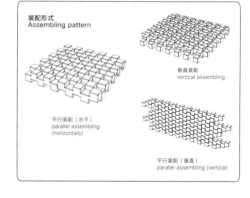

垂直装配
vertical assembling

平行装配（水平）
parallel assembling (horizontally)

平行装配（垂直）
parallel assembling (vertical)

水砖用旋转成型的方法制作而成。

The respective units were made with a method called rotational molding.

堆叠
stacking

网格
grid

Tetchan餐厅
Tetchan

位置：日本东京
设计时间：2014.03—2014.04
建造时间：2014.07—2014.10
主要用途：餐厅
总建筑面积：31.18m²
建筑施工：泷新建设（Takishin）

我们尝试利用废弃的电缆，让家具、灯具、墙壁和天花板变得柔和。覆盖上电缆以后，一切都变成了柔软蓬松的东西。同样，废弃的亚克力经熔化处理后，变成一种有很多气泡的再生材料，表现力很强，我们用它做了吧台椅子。

Discarded LAN cable (which we call "Mojamoja") was recycled and used as furniture, lighting fixtures, walls and ceiling in an attempt to make them soft. All items were reborn as soft and fluffy things by simply attaching LAN cable. Likewise, the rich expressions of air bubbles in recycled products called acrylic balls that were formed by melting discarded acrylic pieces were used as the material for the chairs at the counter.

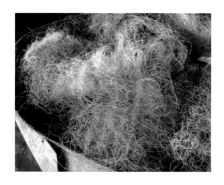

粒子化
particlizing

包装
wrapping

电缆的颜色多而杂，我们利用这一点，设计出一种色彩丰富的云，让人想起印象派的点彩画。

The varied colors of LAN cable was used as an advantage by using the cable to create something that looks like a cloud that evokes an image of pointillism used by impressionists.

网格
grid

泡沫屋
Bubble Wrap

位置：日本大阪
时间：2011.07—2011.08
主要用途：装置

这是一个聚氨酯材料的穹顶装置，顶上嵌套着森万里子的作品《白洞》。先将工地上常用的聚丙烯材料防护网从空中悬挂下来，然后现场喷上聚氨酯泡沫，使穹顶成形。两层穹顶连在一起，获得结构上的稳定性。安东尼奥·高迪在圣家族大教堂中，也尝试过这种用颠倒悬挂来获得合理形态的力学方法，我们在这里用聚氨酯泡沫进行了再现。

A polyurethane pavilion that is suspended in the air from polypropylene curing net that is frequently used at worksites was made to envelop "White Hole," a piece of work by the artist Mariko Mori. The polyurethane foam was sprayed on to create a dome mold, and double domes were connected to achieve structural stability. A mechanical method of achieving a rational form by means of suspension in the opposite direction attempted by Gaudi at the Sagrada Familia was replicated with polyurethane foam.

穹顶是内外两层的结构，内部的穹顶为了遮光，做得很厚；外部的穹顶尽可能做薄，让光线透进来。两层穹顶相互支撑，成为一个没有其他结构部件的纯粹的泡沫装置。

The dome has a two layer structure. Inner one is thicker to block light from outside. Outer one is thinner having porous texture. The each layer support each other without any other reinforcement.

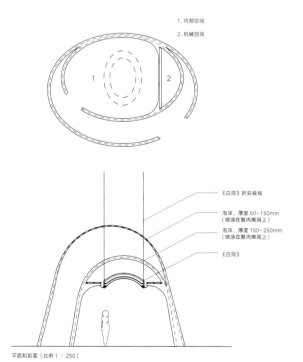

1. 内部空间
2. 机械空间

《白洞》的安装线

泡沫，厚度 50~150mm
（喷涂在聚丙烯网上）

泡沫，厚度 150~250mm
（喷涂在聚丙烯网上）

《白洞》

平面和剖面（比例 1：250）

1.悬挂防护网获得垂曲线。
1.Catenary curve by hanging net.

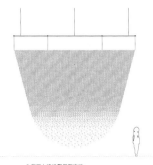

2.在网上喷涂聚氨酯泡沫。
2.Spraying urethane foam to fix the catenary curve.

3.翻转。
3.Rotate.

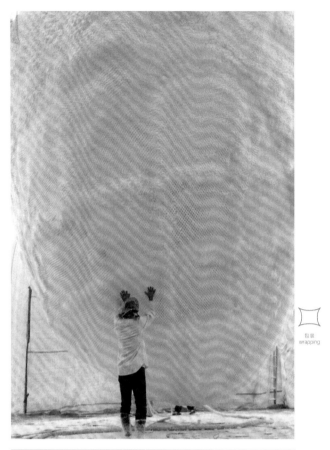

包裹
wrapping

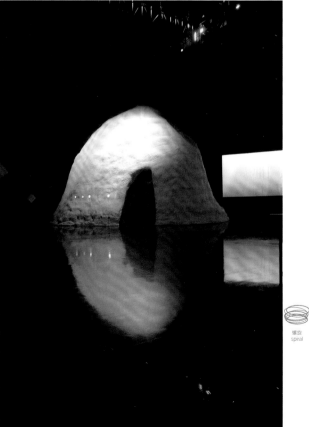

螺旋
spiral

Fire 火　195

北京茶室
Beijing Tea House

位置：中国北京

时间：2010—2014.12

主要用途：茶室、会所（仅限会员）

总建筑面积：250m²

我们把水砖拓展为"框架+面板"的结构，做成能够抵御北京冬季严寒的永久性建筑。以4种旋转成型的方法制造塑料水砖，用其搭建起框架，框架中间嵌入中空的PC板，成为隔热性能高、能够抵御严寒的外墙结构系统。

北京原本是一个青砖砌出来的城市，我们以塑料水砖砌出一个建筑，也是对北京建筑文化传承的一次尝试。

A frame + panel structure that is an evolution of the plastic containers used for the Water Branch House was used to build a permanent building that can withstand the harsh weather in Beijing. Plastic containers made using four types of rotating molds were used as the frame, and hollow panels made from polycarbonate were put in the center. This facilitated the development of an exterior wall structure system with high thermal insulation properties that can withstand winters in Beijing.

Black brick tiles made with reduction firing that were stacked up to build homes and other structures in the past in Beijing. Plastic containers were stacked as small units in an attempt to carry on the history of masonry construction in Beijing.

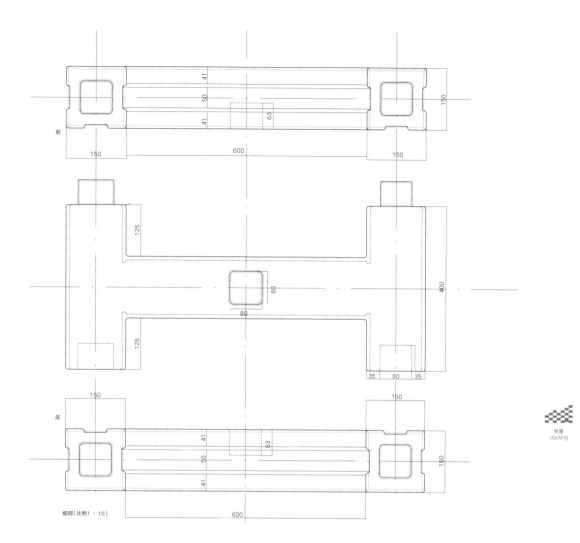

前

150   600   150   150

41
50
63
41
150

125

80
80

400

125

35 80 35

底

150   150

41
63
50
41
150

150

细部（比例1：10）   600

堆叠
stacking

塑料水砖是在上海的工厂制作的。

All plastic containers were made at a factory in Shanghai.

网格
grid

相合家具制作所
家具展示中心
Sogokagu Design Lab

位置：日本三重县
设计时间：2013.02—2014.08
建造时间：2014.09—2015.08
主要用途：写字楼、工厂
占地面积：31,539.79m²
总建筑面积：974.73m²
结构顾问：江尻建筑结构设计事务所
建筑施工：德仓建设株式会社（Tokura Corporation）

我们在北海道草原中做实验住宅"米姆草地"时用到过双层膜，在这里，我们进一步在双层膜中夹入半透明的聚氨酯泡沫，获得了明亮柔和的外墙。我们用的是一种名叫"100倍发泡"（原液100倍发泡）的聚氨酯泡沫，气泡比普通聚氨酯更多，透光性也更好。

在施工现场，将聚氨酯泡沫喷涂到网板上，然后外覆ETFE（聚氟乙烯）膜，内衬PVC（聚氯乙烯）膜。白色的ETFE膜可以阻隔99%的紫外线，防止聚氨酯黄变。

Translucent urethane foam was sandwiched between two membranes, representing an evolution of the two membranes used at Memu Meadows, was used for the bright soft walls of this structure. These walls were created with a special polyurethane foam containing many more air bubbles than ordinary polyurethane called "100 times the air bubbles" (foam expands polyurethane liquid 100 times).

包装
wrapping

The foam was sprayed on using a mesh as the foundation. The exterior members are protected with ETFE and the interior membranes are protected with PVC sheets. The white ETFE membrane cuts out 99% of ultraviolet rays, preventing yellowing of the urethane.

网格
grid

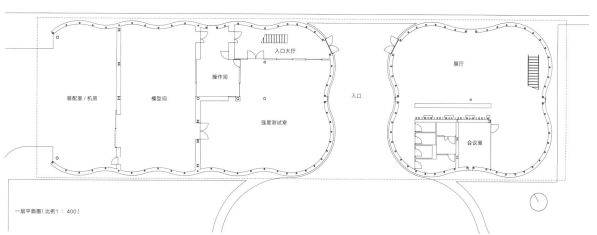

一层平面图（比例1：400）

装配室/机房　模型间　操作间　入口大厅　强度测试室　入口　展厅　会议室

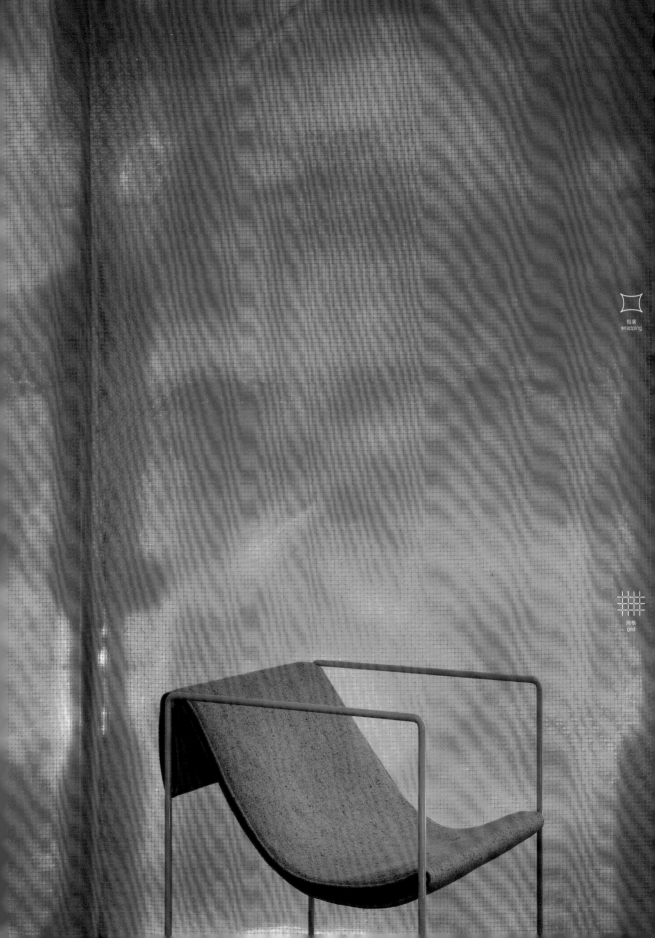

包装
wrapping

网格
grid

金属

Metal

因为石头像混凝土，我有很长时间对它敬而远之。同样，对于20世纪的另一种材料铁，乃至同类的金属，我也曾长期回避它们。不过，一种能够恢复形状的"形状记忆合金"改变了我的想法。我开始感受到金属的友好，开展了各种关于金属的实验。

我在庆应义塾大学执教的时候，参加了一场很独特的硕士论文发表会，是关于用形状记忆合金做钢架连接件的研究。地震时这种连接件即使变形，加热后也会"想起"原来的样子，恢复原形。虽然是金属，但它简直就像有生命的东西。实际上，形状记忆合金常用在衬衫领子、女性文胸等贴身衣物上，因为能随着体温变形，对人体非常友好。有这么精细、友好的金属，我就想用它做一个建筑上的实验。2005年，我做了KXK，一个能随着温度变化改变形状的装置。

做这个装置的关键，是要让整个结构尽可能地弱，如果太强，温度变了形状也不会变。可是太弱了也不行，结构立不起来，作为建筑是不成立的。不能太强也不能太弱，这成了我寻求新建筑方向的一次尝试。

在强与弱之间实现微妙的平衡，金属是最合适的材料。因为金属不仅仅是强硬的，还是有韧性的。

利用金属韧性最好的例子就是金属拉网——

Like stone, which I long avoided because it resembles concrete, metal is another main 20th-century material that I have long avoided. But the encounter with a remarkable shape-memory alloy (SMA) changed my mind, and now I am starting to think of metal as a friend and to use it in various experiments.

At the time I was teaching at Keio University, and there was a master thesis presentation for a unique research project. It involved magnesium SMA joints for rigid-frame steel structures. Even if the joints were deformed by an earthquake, they would recall their shape when heated and return to their original state.

Even though it was metal, this seemed like a living thing. In everyday life, SMAs are used in items of clothing close to the body, like collars and brassieres that utilize body heat for shape changes. I thought that since such a delicate and supple metal exists, I should design something with it. The result was KXK (2005), a pavilion that changed shape with heat.

In this design, the critical point was to make the architecture as weak as possible, because if it was too strong then it would not change shape when the temperature changed. Of course, it could not be too weak either, because if it could not stand on its own then it would not be architecture. This project was one step toward a new direction in architecture: not too strong, and not too weak.

Metal is the best material to achieve a delicate balance between strength and weakness. It is not only strong, but also resilient.

The material that takes best advantage of metal's

在金属板上做出切缝，然后拉伸成网状。这种金属拉网上很容易附着东西，我曾尝试附着纸浆和泥土。这样，我对金属的实践就和土、纸联系到一起了。

resilience is expanded metal—metal that has been cut and stretched to form a mesh. Various other materials can be easily attached to the mesh. We have tried attaching paper and earth, and have found that the system diagrams for metal, paper, and earth come together into one.

1

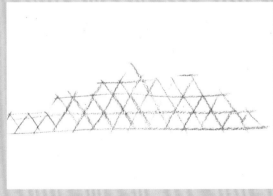

2

Sketches by Kengo Kuma
1. KXK
2. Polygonium
3. Green Cast
4. Darius Milhaud Conservatory of Music

隈研吾手稿
1.KXK
2.铝板拼接陈列架
3.Green Cast 综合建筑
4.达律斯·米约音乐学院

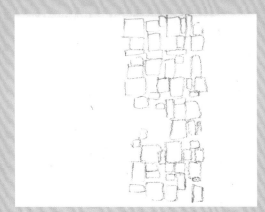

3

4

KXK
KXK

位置：日本东京
时间：2005
主要用途：装置
总建筑面积：5m²
结构顾问：橡树结构设计事务所

我们用直径2mm的形状记忆合金（硬度随温度变化而变化的镁合金）做了一个球体装置，它像生物一样，形态会随着温度的变化产生柔和的变化。

先用合金做成直径为300mm的圆环，圆环与圆环之间用塑料绑带连接，形成一个球形的结构体。合金很柔软，不能自立，需要先用泡沫做一个球形的模型，把合金圆环贴附在上面，形成球体后再把泡沫取出来。合金结构体上覆盖一层半透明的EVA（乙烯-醋酸乙烯共聚物）网，看起来如蚕茧般柔和。这些圆环拆解后可以装进一个小盒子，搬运很方便。

Shape-memory alloy (magnesium alloy that changes hardness depending upon temperature) with a diameter of 2 mm was used to build a dome like an organism that softly changes its shape when the temperature changes.

First, 300 mm diameter rings were made with the alloy, and a dome was formed by connecting the rings with plastic insulation locks. Because the rings are soft and not stand, a dome mold was first created with styrofoam, and the styrofoam was removed after completing the dome by attaching the rings. The structure was covered with a semitransparent mesh called EVA sheet in order to give it the impression of being soft like a cocoon. The rings fit in a small box when disassembled, making it easy to transport them.

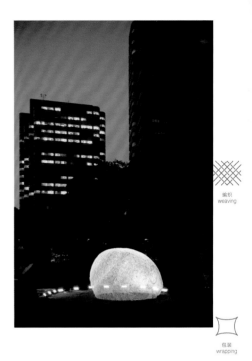

编织
weaving

包裹
wrapping

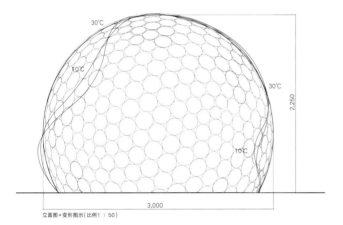

立面图+变形图示（比例1：50）

螺旋
spiral

铝板拼接陈列架
Polygonium

位置: 日本富山县

时间: 2008年9月起，每年秋季

主要用途: 家具陈列

我们在实验装置"石片城堡"中做过以三角形为单元的立体格子，在这里，我们用挤压成型的铝板（10mm厚）代替了原先的石板。利用挤压铸造的特点，我们把连接部位做成严密嵌合的样式。不用螺丝，只需打入一个销子，就能把面板连接起来，拆解、添加也都能在短时间内完成。蒲由奈还用这种结构做过一个实验住宅（211页上部照片），当时他是富山大学的学生。

The cubic lattices that use triangles as the unit which were pursued with Stone Card Castle were applied to extruded aluminum sandwich panels (10 mm thick). The characteristics of extrusion were utilized to make the joints the same shape as a seal case. The objective was to create a system in which panels can be connected to each other by simply tapping in a pin and not using bolts or screws, and can be disassembled or added to in a short amount of time. Experimental housing using this system is a piece of work by Yuna Kaba, a student of Toyama University (at the time).

强化玻璃，厚度10mm，锥形边缘，贴膜

防滑硅胶垫，厚度1mm

连接件: 挤压铝

铝面板氧化铝涂层透明涂料，厚度10mm

快速别针

玻璃连接件细部（比例1：1）

连接件: 挤压铝

快速别针

铝面板氧化铝涂层透明涂料，厚度10mm

10

铝板连接件细部（比例1：1）

在以金属铸造闻名的富山县高冈市，古老街市的质朴面貌与陈列架的精致细节相得益彰。

The delicate detail of Polygonium blends in with the delicate appearance of the old townscape in Takaoka, Toyama, known as a major town of producing cast-metal objects.

支撑
reciprocating

堆叠
stacking

多边形
polygon

Green Cast 综合建筑
Green Cast

位置：日本静冈市

设计时间：2009.02—2009.09

建造时间：2010.05—2011.06

主要用途：居住、办公、诊所、药店

占地面积：424.5m²

总建筑面积：1,047.8m²

结构顾问：牧野结构设计

建筑施工：竹中工务店（Takenaka Corporation）

用12mm厚、拥有自然纹理的铸铝板做外墙装饰构件，把通风口、雨水管、绿化箱、绿化基材和排水管等全部整合起来。制作铝板时，先在泡沫塑料上滴胶，形成腐蚀图案，再将纹理转印到铝板上，使铝板获得岩石般的肌理。铝板可以旋转、倾斜，形成生机勃勃的绿色外墙，看起来自然随意。

Cast aluminum with an organic texture that is 12 mm thick was used to design exterior wall units that integrate air vents, rainwater drainpipes, planter boxes, greening substrates and drainpipes. Patterns in handmade molds created by dripping adhesive on styrofoam and causing corrosion were transferred to the surface of the aluminum die cast pieces, making the surface look like rock. The same unit can be rotated or inclined, achieving an organic green exterior wall that does not appear to have any regularity.

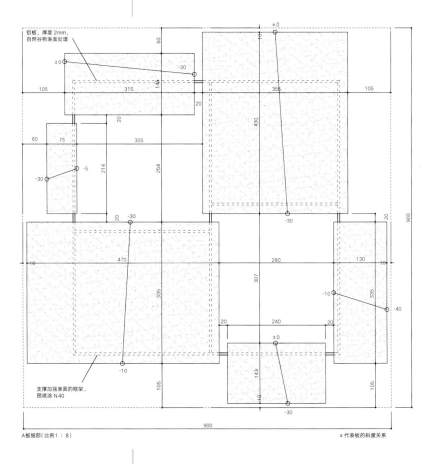

铝板，厚度 2mm，自然谷物表面处理

支撑加强表面的框架，预喷涂 N40

A板细部（比例 1：8）

± 代表板的斜度关系

粒子化
particlizing

包装
wrapping

网格
grid

达律斯·米约音乐学院
Darius Milhaud Conservatory of Music

位置：法国普罗旺斯艾克斯

设计时间：2010—2011

建造时间：2011—2013

主要用途：音乐、舞蹈和戏剧学院

占地面积：2,045m²

总建筑面积：7,430m²

结构顾问：Cer3i公司（Cer3i）

建筑施工：马蒂厄·法柳（Mathieu Faliu）、米鲁纳·康斯坦丁内斯库（Miruna Constantinescu）

我们用铝板做出折纸般的外立面，建筑的开口藏在微妙的褶皱里，墙与开口在褶皱里融为一体。我们用的是弯板，着重表现铝板的光泽和质感。所有铝板的厚度都是4mm，板缝最窄处仅为5mm，这样才有了与普通铝质外立面完全不同的感觉。

粒子化
particlizing

褶皱根据不同方位有微妙的变化，同时也能遮挡法国南部强烈的阳光。出生于普罗旺斯艾克斯的音乐家达律斯·米约，其音乐硬质而自由，建筑硬质的褶皱正与之呼应。

包装
wrapping

Aluminum panels were used to create a facade that looks like origami, and openings were made in the subtle creases in the facade to enable the wall and the openings to be merged into the creases. Care was taken to not detract from the hard brilliance and texture of the aluminum when using the bent plates. All cut plates had a thickness of 4 mm, and the minimum spacing between plate joints was 5 mm, giving a completely different impression from an ordinary aluminum facade.

The creases differ in a subtle manner depending upon the orientation, serving the role of preventing the strong sunlight in Southern France from coming inside. The rigid yet free music made by the musician Darius Milhaud from this town (Aix-en-Provence) for which this facility is named will reverberate against these rigid creases.

网格
grid

多边形
polygon

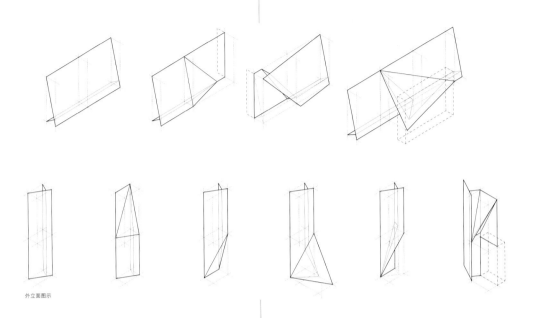

外立面图示

粒子化
particlizing

包装
wrapping

网格
grid

多边形
polygon

无锡万科
Wuxi Vanke

位置：中国无锡
设计时间：2010.04—2012.06
建造时间：2012.07—2014.04
主要用途：美术馆、写字楼、商场
占地面积：6,217.92m²
总建筑面积：10,440.22m²
结构顾问：结构网（扬原重雄）[Structural Net
（Shigeo Agehara）]

继Green Cast的铸铝板之后，我们进一步设计了另一种天然有机形态的多孔质铸铝板，看上去有点像无锡太湖出产的太湖石。铝板表面凹凸起伏，灵感来自曲面层叠形成的等高线。纵向用直径48mm的不锈钢管，横向用铸铝板，编织成灵活的曲面，能够根据内部设施的规划而调整。这样的外立面与单纯用铁板或铝板做成的不同，有一种柔和的质感。

The aluminum cast panels used in Green Cast were further refined to design porous cast aluminum units that will cause people to recollect Taihu and other stones from Taihu Lake near the planned site. A hint for the cast shape was obtained from an overlapping contour maps of the surface. The exterior wall which was woven using 48-mm-diameter stainless steel pipes as the warp and aluminum cast units as the woof created a soft texture that is completely different from steel and aluminum panels, and create free curves according to the program inside the facility.

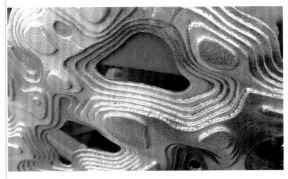

粒子化
particlizing

包装
wrapping

这是一个砖砌老厂房的增改建项目，我们将其改建成了一个美术馆。

A museum was planned as an expansion of an existing mill that was built with brick.

网格
gnd

螺旋
spiral

铝云
Le Nuage d'Aluminium

位置：法国巴黎
设计时间：2016.12—2017.11
建造时间：2017.10—2017.11
主要用途：装置
尺寸：500m²
顾问：特科纳（TECHNAL）

我们与法国的铝制品厂家特科纳合作，根据巴克敏斯特·富勒基于一种基本结构体系提出的"张拉整体式结构"理论（结构同时具有张力和拉力），开发了一种空间构件。

以挤压成型的直径40mm的铝管为压杆，配合直径2.5mm的不锈钢拉索，构建出墙和屋顶，仿佛金属做成的云。桌椅也沿用了同样的结构体系。铝管随意地反射着光线，仿佛水蒸气粒子聚合而成的云，给人置身云雾般的空间体验。

A space unit with a tensegrity structure (combination of tension and integrity) based on an ultimate structure system proposed by Buckminster Fuller was developed in collaboration with Technal, a manufacturer of aluminum products in France.

Extruded 40-mm-diameter aluminum pipe compression members and stainless steel 2.5-mm-diameter cable expansion members are assembled to create walls and ceilings with metal that look like clouds. Tables and chairs are also assembled with the same system. Light that is reflected in a random manner by the aluminum pipes makes it possible to have the same type of spatial experience as clouds which are a collection of water vapor particles.

桌椅的结构体沿用了张拉式整体结构。

The tensegrity system was expanded as the structure for tables and chairs.

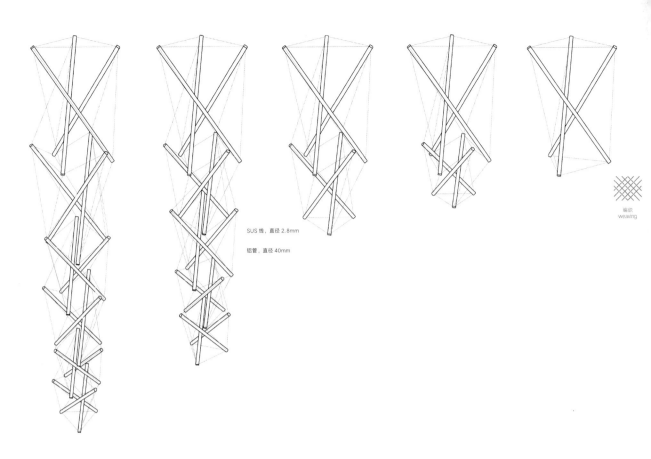

SUS 线，直径 2.8mm

铝管，直径 40mm

编织
weaving

多边形
polygon

虹口 SOHO
Hongkou SOHO

位置：中国上海
设计时间：2011.11—2013.08
建造时间：2012.04—2015.10
主要用途：写字楼、公共办公空间、零售店
占地面积：16,000m²
总建筑面积：95,000m²
结构顾问：江尻建筑结构设计事务所、
英海特（Inhabit）（幕墙）

用18mm宽的铝条编织成高透明度的网片，用在高层写字楼的外立面上，给大厦带来一种波动起伏的感觉，建筑仿佛将要消失在空中。

将铝网片做出褶皱，褶皱形成的投影和阴影随着高度的变化而改变，在不同的天气、时间和季节也各不相同。为了耐腐蚀,铝网片背后的底板也都是铝材。

Highly transparent mesh that was woven using 18 mm wide aluminum plates was used as the exterior material for a high-rise building in order to create a shimmering surface with an expression that tends to disappear into the sky.

The aluminum mesh is bent into pleats, which change shade and shadows according to the height of the pleats, creating a facade that shimmers in a variety of ways depending upon the weather, time and season. All material beneath the panels is aluminum as a consideration for possible corrosion.

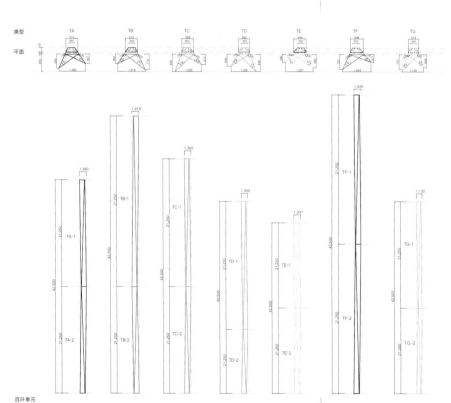

百叶单元

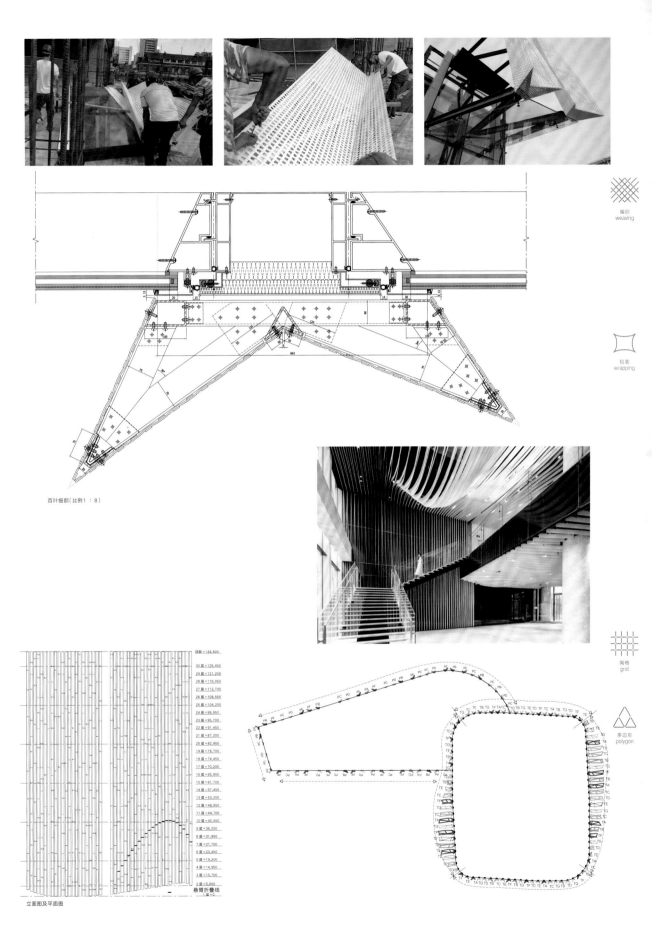

百叶细部（比例 1 : 8）

编织
weaving

包裹
wrapping

网格
grid

多边形
polygon

立面图及平面图

顶部 =133.500
30 层 =125.450
29 层 =121.200
28 层 =116.950
27 层 =112.700
26 层 =108.550
25 层 =104.200
24 层 =99.950
23 层 =95.700
22 层 =91.450
21 层 =87.200
20 层 =82.950
19 层 =78.700
18 层 =74.450
17 层 =70.200
16 层 =65.950
15 层 =61.700
14 层 =57.450
13 层 =53.200
12 层 =48.950
11 层 =44.700
10 层 =40.450
9 层 =36.200
8 层 =31.950
7 层 =27.700
6 层 =23.450
5 层 =19.200
4 层 =14.950
3 层 =10.700
2 层 =5.600
悬臂折叠线
1 层 =0

建筑的幕墙从上延伸到下，使大厦看上去仿佛是大地的延伸。

The curtain wall detail of the high-rise portion of the building was extended to the low-rise portion in order to achieve a high-rise building that appears to be an extension of the earth.

北京前门街区
Beijing Qianmen

位置：中国北京
设计时间：2015.03—2015.09
建造时间：2015.09—2016.11
主要用途：办公室、咖啡店
总建筑面积：393m²

这是北京前门广场东侧的一个街区改造项目。前门一带还保留着由胡同和四合院构成的老北京风貌，我们需要做的是，在尽量保留老建筑的同时，让这一带变成一个艺术与设计并重的街区，焕发新的活力。

我们把四合院完全封闭的青砖墙打开了一部分，用挤压成型的铝质构件组合成幕墙，形成了一个面向胡同开放的新型低层街区。我们希望铝质的幕墙既能成为一种"透明的砖"，也能成为遮阳的屋檐，具有多种功能。这种构件组合可以自由增减，能够灵活应对未来街区的变化。

This project was implemented to rejuvenate the old part of Beijing called Qianmen (area on south side of Tian' anmen Square) that retains the most of its historical characteristics consisting of Hutong (alleys) and Siheyuan (courtyard houses) as an area with a focus on art and design, while preserving the old architecture.

Portions of the walls of the Siheyuan houses that were completely closed off with black brick tiles called Sen were replaced with screens made by assembling extruded aluminum units to create a new area of low-rise buildings in Beijing that is open to the Hutong alleys. The design objective for the aluminum screens was to make them "transparent bricks" while the eaves that provide shade during the day are provided with multiple functions. The assembled extruded aluminum units can be freely added to or subtracted from, facilitating a flexible response to future changes in the city.

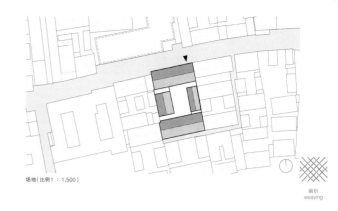

塔地（比例1：1,500）

编织
weaving

包装
wrapping

网格
grid

四合院的房屋看起来像是青砖垒砌的砖石结构建筑，实际上是由木柱和木梁构成的框架结构。因为是框架结构，所以砖可以替换成铝质幕墙。

Siheyuan houses which look like black brick masonry structures actually have a wood column-beam frame structure. This frame structure enabled the bricks to be replaced with aluminum screens.

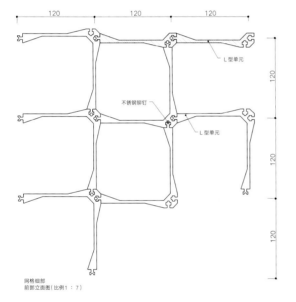

网格细部
前部立面图 (比例 1 : 7)

阳澄湖游客集散中心
Yangcheng Lake Tourist Transportation Center

位置：中国苏州
设计时间：2012.09—2016.09
建造时间：2018.05
主要用途：游客中心、码头
总建筑面积：7,759.34m²

屋顶上随机铺排着200×100mm规格的L型角铝，有着茅草屋顶般的毛糙质感。

阳澄湖以出产大闸蟹闻名。我们希望闪烁着不规则光泽的铝屋顶，能够与阳澄湖清澈的水面相映成趣。

粒子化
particlizing

The roof was made by randomly arranging L-200 × 100 mm aluminum angles in order to bring back the shaggy deep expression of a thatched roof.

The main objective was to create an effect where the aluminum roof that shines in a random manner blends in harmoniously with the beautiful surface of Yangcheng Lake, which is known for its Chinese mitten crab.

包装
wrapping

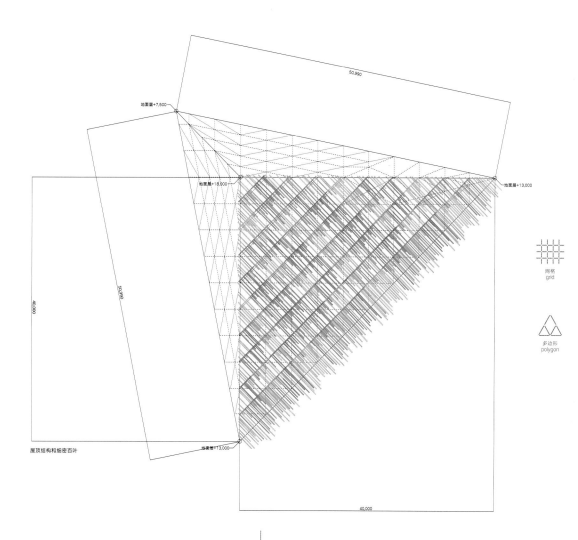

网格
grid

多边形
polygon

为了获得自然天成的效果，我们把挤压成型的铝材的顶端切割成不规则形态。

The edges of the extruded aluminum were cut in a random manner in order to provide randomness similar to that found in nature.

粒子化
particlizing

包装
wrapping

网格
grid

多边形
polygon

膜／纤维

Membrane / Fiber

如果想让建筑变得更加柔软，最有趣、最有可能用的材料就是纤维。戈特弗里德·森佩尔认为，建筑是"编织"出来的。最纯粹的编织行为产生的就是布，我现在最感兴趣的就是像布一样柔软蓬松的建筑。

观察生物，你可以发现生物的身体原本就是由重叠的纤维构成的。人也是这样，多层柔软的纤维构成人的身体，身体外面又覆盖着衣服的纤维层，再往外，还有建筑的纤维层。如果建筑能变得忽大忽小，就像穿上和脱下的衣服那样，那么一定很有趣。

就像日本传统礼服"十二单衣"（穿着多层衣服，起源于平安时代），日本有把多层单薄的纤维布料重叠起来使用的传统。用这种手法，建筑与身体的对立、建筑与景观的对立，乃至人工与自然的对立，都会消失，环境中的一切都能重叠、融合。我们怀着这样的梦想，去挑战纤维的建筑。

随着对纤维的了解越来越多，我们发现纤维不限于简单的织法，它有很多不同的结构体系，不仅仅是经线、纬线两个轴向的编织，还可以是相互呈120°角的三个轴向的编织。这种三轴编织能够做出三维立体的织物，我们曾经把它用在室内设计上。

在金属结构体上覆盖纤维是很多人都想得到的，其实纤维本身就可以内藏结构，是可以自立的。

If the goal is to make architecture as supple as possible, then the most interesting material and the one with the most potential is fiber. Semper realized that architecture is made through a process of weaving. When the act of weaving is carried out in its purest form the result is cloth. Right now what interests me the most is architecture that is soft and supple like cloth.

Observation shows that the bodies of animals are made up of overlapping layers of fibers. The body itself is made up of many soft layers, and then it is covered by further layers—clothes. Architecture as fiber adds even more layers. I think it would be interesting if architecture could grow or shrink like clothes being worn or taken off.

As shown by the aristocratic "twelve-layer robe," Japanese fashion traditionally emphasized thin layers of cloth. Taking that approach seriously could dissolve the oppositions between architecture and the body, between architecture and landscape, and between the natural and the artificial. Everything in the environment will overlap and commingle. Toward that vision, we are working on the challenge of architecture as fiber.

As we learned more about fibers, we discovered various structural systems that exists while they are not yet woven. We also discovered triaxial weaving. Most fabrics are biaxial, with X and Y axes. Triaxial weaving has three axes, displaced by 120 degrees each. We found that triaxial weaving makes it easier to create 3D fabrics, and used it as an interior material.

The idea of draping fibers over a metal structure might occur to anyone. But fibers themselves can have

1

2

"软"和"硬"的概念也是相对的，不需要把它们对立起来。纤维通常意味着软和弱，我们却尝试用它去提高又硬又重的混凝土框架结构的抗震性能。20世纪又重又硬的物质是脆弱的，现在我们可以用柔软的物质去拯救它。

structure and be independent elements.

The concepts of "soft" and "hard" are relative and no longer need to be opposed. We have worked on the challenge of using fibers, normally soft and weak, to enhance the earthquake resistance of rigid frames in hard and heavy concrete. A contemporary soft material compensates for the vulnerabilities of a hard and heavy 20th-century material.

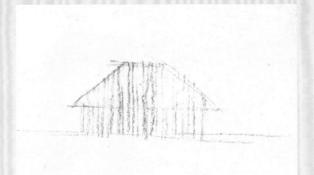

3

4

福崎空中广场
Fukuzaki Hanging Garden

位置：日本大阪

设计时间：2004.01—2004.07

建造时间：2004.08—2005

主要用途：写字楼、广场

占地面积：2,450m²

总建筑面积：981m²

结构顾问：橡树结构设计事务所

建筑施工：三和建设公司

（Sanwa Kensetsu Corporation）

项目位于海边的人工填埋地。作为一座面临搬迁的临时建筑，需要轻质、可重建的外墙结构。我们把目光投向了经常作为门帘、隔断，出现在仓库和工厂的PVC软片。

PVC软片不必嵌在门窗框中，只用不锈钢丝缝合，便能满足建筑所需的密封性。两层金属拉网叠在一起，纹路呈90°，成为透明又具有抗震性能的墙体。

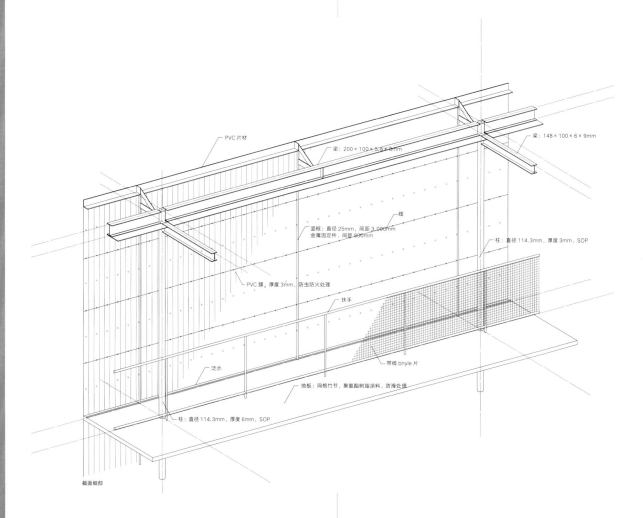

PVC 片材

梁：200×100×5.5×8mm

梁：148×100×6×9mm

线

竖框：直径 25mm，间距 3,000mm
金属固定件，间距 800mm

柱：直径 114.3mm，厚度 3mm，SOP

PVC 膜，厚度 3mm，防虫防火处理

扶手

带线 bnyle 片

泛水

地板：网格竹节，聚氨酯树脂涂料，防滑处理

柱：直径 114.3mm，厚度 6mm，SOP

截面细部

There was a need for a light reusable cladding system that is appropriate for temporary structures on reclaimed land that are going to be moved, and we focused on soft polyvinyl chloride curtains that are used as partitions that resemble shop curtains at the entrances to warehouses and plants.

Rather than being fixed to a sash or frame, these vinyl curtains are simply sewed to a stainless steel wire, providing the required level of airtightness for a building. Double layers of expanded metal were bonded together with its pattern crisscrossed, in order to achieve a transparent earthquake-resistant wall.

包裝
wrapping

网格
grid

茶室
Tee Haus

位置：德国法兰克福

时间：2005—2007

主要用途：茶室

总建筑面积：31m²

薄膜结构：坎诺比奥

建筑施工：竹中工务店（欧洲）（地基）

顾问：TL公司（TL）的结构顾问师、太阳铁工（欧洲）（Taiyo Europe）

这个临时的便捷式茶室充气后才能使用，因此需要用多次膨胀也不会损坏的膜来制作。我们采用了一种以PTFE（聚四氟乙烯）为原料的新材料"特德拉"。

茶室为双层膜结构，膜与膜之间可充入空气。用涤纶绳子将两层膜系紧，绳与绳交错形成长宽各600mm的网格，以至于外表看起来像毯子一样。茶室的入口装有防水拉链。

A new material called Tetra that uses PTFE (Polytetrafluoroethylene) as the raw material was used for a tea house that can be instantly inflated with air when it is used, that needed to be made with a membrane that would not deteriorate no matter how many times it was inflated.

It has a double membrane structure that can be inflated by filling it with air, and the double membrane is bonded together with polyester cord in a 600 mm grid to give the surface the look of a quilt. The miniature tea room entrance is provided with a waterproof zipper.

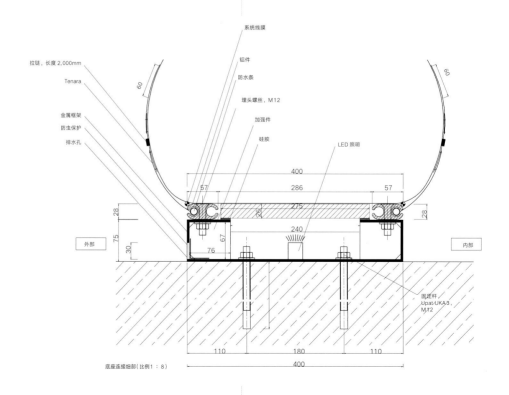

底座连接细部（比例1：8）

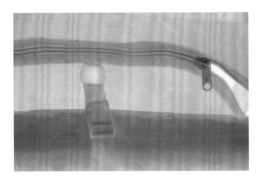

包装
wrapping

网格
grid

Membrane / Fiber 膜 / 纤维 241

浮庵
Floating Tea House

位置：日本
时间：2007
主要用途：茶室
总建筑面积：7.29m²
顾问：NUNO公司（NUNO）、井上工业（Inoue Industries）、D.P.巴隆（D.P.Balloon）

给PVC材料的气球充入氦气使之飘浮，在其上盖上一层名叫"超级欧根纱"的涤纶纤维透明纱——据说这是世界上最轻的面料。这样一个茶室，设计上需要考虑氦气的浮力与纱重量之间的平衡。茶室可拆卸搬运，曾被装在行李箱里从日本带去华盛顿展出。

A polyvinyl chloride balloon was filled with helium gas to make it float, and it was covered with a cloth called Super Organza made from polyester fibers which is the lightest fabric in the world in order to create a floating tea house. It was designed so that the buoyancy of the helium is balanced with the weight of the Super Organza, and is mobile. It was placed in a suitcase and taken from Japan to Washington D.C. for the exhibition.

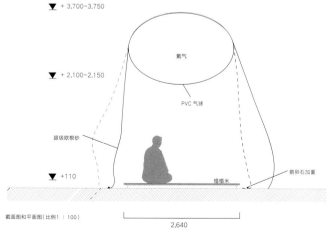

截面图和平面图（比例1：100）

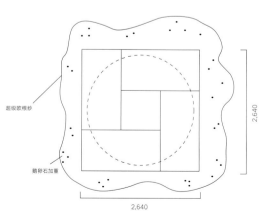

包装
wrapping

螺旋
spiral

伞屋
Casa Umbrella

位置：意大利米兰

时间：2007—2008

主要用途：房屋原型

总建筑面积：15m²

结构顾问：江尻建筑结构设计事务所

伞形结构顾问：IIDA KASATEN伞店（IIDA KASATEN）

织物顾问：杜邦建筑创新与杜邦科技（DuPont Building Innovation & DuPont Technical Center）

巴克敏斯特·富勒证明了由15个面组成的"富勒圆顶"的合理性。基于此，我们做了一个牧民帐篷般的实验装置：让大家撑起伞，用15个防水拉链把伞连接起来。装置可以自由组装或拆卸。伞面材料采用一种廉价的无纺布"特卫强"。打开一个三角形的区域，就有了通风的窗口。伞面作为受拉件，伞骨作为受压件，成为一个张拉整体式结构。比起富勒的圆顶，我们的结构框架要轻巧得多。

Buckminster Fuller proved the rationality of the Fuller Dome that is made with 15 faces. Based on this concept, we created the ultimate nomad shelter that can be freely assembled and disassembled by having people bring umbrellas and connecting them together with 15 waterproof zippers. A reasonable polyester waterproof sheet called "Tyvek" was used as the material for the umbrellas, and triangular pieces were added to create ventilation windows that can be opened and closed. The membranes function as the tension members, and the thin umbrella ribs function as the compression members, creating a type of tensegrity structure, achieving a frame that is much thinner and lighter than the Fuller Dome.

× 15 =

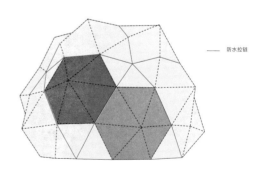

防水拉链

防水拉链

日常用法

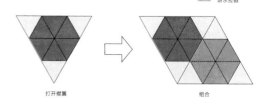

打开襟翼

组合

支撑
reciprocating

包装
wrapping

堆叠
stacking

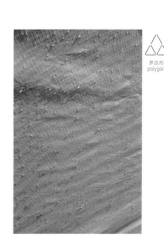

多边形
polygon

空气砖
Air Brick

位置：中国上海
设计时间：2010.01—2010.07
建造时间：2010.07—2010.08
主要用途：装置
占地面积：100m²（展览区域）
总建筑面积：30m²
结构顾问：江尻建筑结构设计事务所

ETFE作为一种透明的膜材很受关注，我们用它开发了一种充气枕头般的单元构件，以构建一个建筑系统。这种ETFE膜加工难度很高，通常只用在大型建筑的屋顶上，我们挑战了ETFE的立体切割及热封技术，做成了这个由小单元连接起来的结构。ETFE的热封通常要与金属框架结合，单纯的ETFE热封世界上还没有先例。各个小单元之间由管道连接，根据需要用空气泵进行充气或放气，使之膨胀或收缩。利用膜，我们做成了一个能像生命体一样呼吸的建筑。用同样的方法，我们做了椅子和桌子。

ETFE which has received considerable attention as a transparent membrane was used to develop a unit building system that is based on a unit that is like an air cushion. We took on the challenge of 3D cutting and heat sealing of ETFE which is considered to be very difficult which is why it has only been used for the roofs on large-scale structures up until now, and achieved a structure that is made by connecting these small units. ETFE is generally based on combined use with a metal frame and heat sealing, and there is no precedent anywhere in the world in which connections have been made with only heat sealing. The respective units are connected to tubes, enabling an air compressor to be used to inflate or deflate the units as desired. Thus, we used a membrane to achieve architecture that continues to breathe like an organism. The same system was used to make the chairs and tables.

支撑
reciprocating

堆叠
stacking

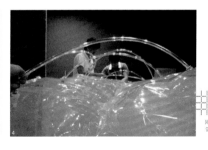

网格
grid

米姆草地·实验住宅
Memu Meadows

位置: 日本北海道

设计时间: 2009.03—2010.10

建造时间: 2010.11—2012.06

主要用途: 居住和其他功能

占地面积: 79.5m²

总建筑面积: 79.5m²

结构顾问: 森部康（Yasushi Moribe）

建筑施工: 高桥工务店（Takahashi Komuten）

这个半透明的实验住宅采用了双层膜——木结构框架的外部覆盖一层聚酯膜，内部再贴一层玻璃纤维膜，让热空气在两层膜之间循环。膜本身不是隔热材料，但两层膜之间的空气层，使这座位于寒冷地区的住宅建筑具备了较高的保温性能。

内部的玻璃纤维膜用双面胶粘贴，容易拆解，方便后续维护。白天阳光透进来，室内不需要照明就很明亮，节能环保。北海道当地阿伊努族的传统住宅有着厚厚的松软外墙，这给了我们设计的灵感。

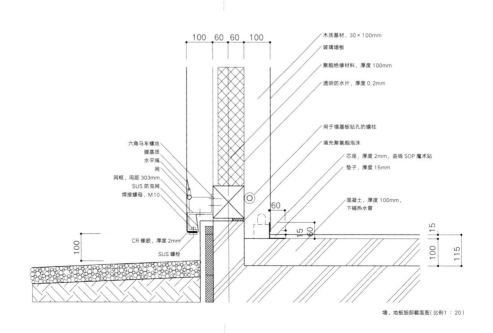

墙、地板细部截面图（比例1：20）

This is a translucent experimental residence that is covered with a polyester membrane. A wooden frame was covered with a polyester membrane that serves as the outer skin, a double membrane was created by attaching fiberglass sheets that serve as the inner skin, and a heater was placed in between the membranes to circulate warm air between them, enabling a cold region residence to be built that has a high level of thermal insulation which uses air in place of heat insulation materials.

The fiberglass membrane that serves as the interior wall has double-sided tape that allows the wall to be attached and detached, making it easy to perform maintenance. The residence is sustainable in that since no insulation is used, sunlight shines in, eliminating the necessity for lights during the day. A hint for the thick fluffy exterior walls was obtained from the Chise traditional Ainu houses.

覆盖屋顶的聚酯膜在工厂里缝制完成后，用起重机盖在木框架上。

The polyester membrane that covers the roof was sewn at a factory, and a crane was used to cover the wood frame with the membrane.

包装
wrapping

网格
grid

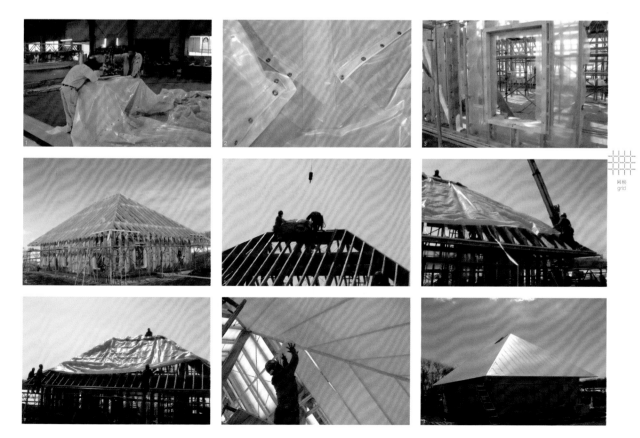

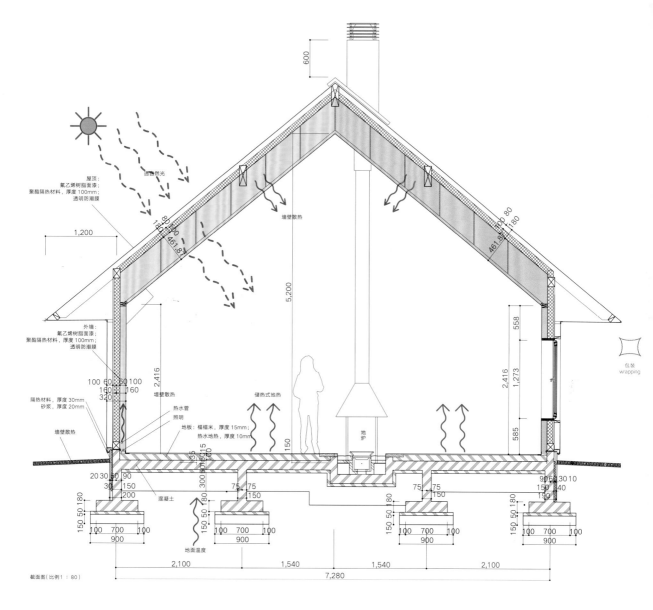

600

1,200

透自然光

屋顶：
氟乙烯树脂面漆；
聚酯隔热材料，厚度100mm；
透明防潮膜

墙壁散热

118.8
461.8

461.8

130.80
180

461.8

5,200

外墙：
氟乙烯树脂面漆；
聚酯隔热材料，厚度100mm；
透明防潮膜

558

2,416

隔热材料，厚度30mm；
砂浆，厚度20mm

100 60 50 100
160   160
320

2,416

墙壁散热

储热式地热

1,273

墙壁散热

热水管
照明
地板：榻榻米，厚度15mm；
热水地热，厚度10mm

150

地炉

585

20 30 50 90
30   150
200

混凝土

35
300 50 15 15
150

75 75

150

地面温度

75 75

150

90 65 30 10
150 40

190

150 50 180

100 700 100
900

150 50 180

100 700 100
900

150 50 180

100 700 100
900

150 50 180

100 700 100
900

2,100

1,540

1,540

2,100

截面图（比例1：80）

7,280

包装
wrapping

网格
grid

"上下"品牌上海旗舰店
Shang Xia Shanghai

位置：中国上海

设计时间：2009.08—2009.12

建造时间：2009.12—2010.06

主要用途：商店

总建筑面积：126m²

结构顾问：江尻建筑结构设计事务所

建筑施工：乃村（北京）（NOMURA Beijing）

我们用立体的六角形织物创造了一个柔和的洞窟般的空间。制作时没有采用通常那种经线、纬线两个轴向的织法，而是先将三根线按照相互呈120°角的三个轴向织成涤纶纤维布片，再用金属模具施压成型。用线将这些半透明的立体布片缝合起来，产生一种自由的形态，就像凹凸起伏的洞窟一样。有着无数缝隙和起伏的"松散的空间"，会让人和其他生物感觉舒服自在。

Special hexagon shaped cloth that was formed into a 3D shape was used to create space that is like a soft cave. Normal biaxial weave cloth that is woven with vertical threads and horizontal threads was replaced with triaxial weave polyester cloth in which the three axial threads are each shifted 120 degrees, and is pressed into shape in a mold which is called 3D shaping. The undulating semitransparent pieces are sewn together, enabling a shape to be freely created that resembles a cave because of the scraggy surface. This "loose system" that has an infinite number of crevices and undulations creates a space that is comfortable for people and other living things.

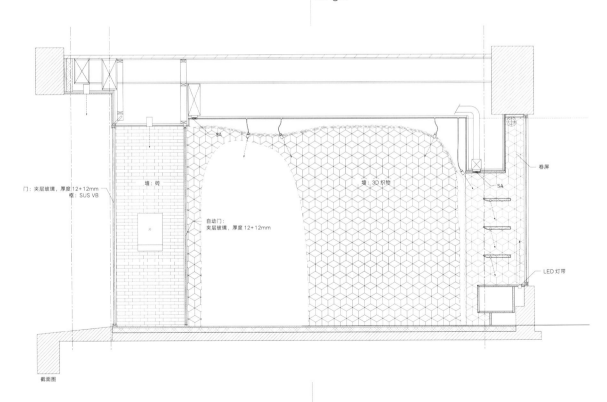

门：夹层玻璃，厚度12+12mm
框：SUS VB

墙：砖

自动门：
夹层玻璃，厚度12+12mm

墙：3D织物

卷屏

SA

LED 灯带

截面图

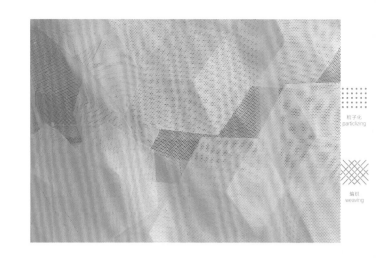

粒子化
particlizing

编织
weaving

包装
wrapping

多边形
polygon

比起含有 90° 角的几何形状，六角形拥有不同的几何性质，可以营造更自由松散的形态。

The space was provided with hexagonal geometry in order to facilitate a loose natural shape that differs from 90 degree geometry.

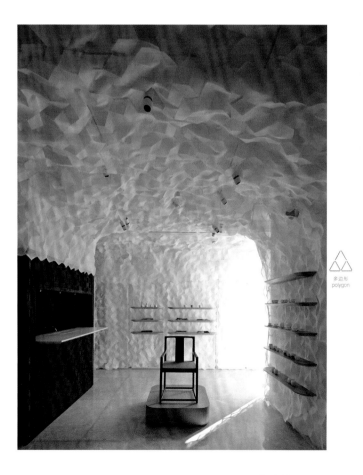

小松精练纤维
研究所
Komatsu Seiren Fabric
Laboratory fa-bo

位置：日本石川县
设计时间：2013.08—2014.11
建造时间：2015.02—2015.11
主要用途：办公室、展示空间
占地面积：121,485.30m²
总建筑面积：2,873.42m²
结构顾问：江尻建筑结构设计事务所
建筑施工：清水建设（Shimizu Corporation）

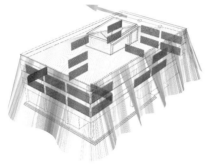

棒状线材系统地震模拟受力分析。
System of rod tensile force bearing at the time of earthquake.

我们利用碳纤维去加强现有混凝土办公楼的抗震性能。碳纤维材料通常只有坚硬的棒状线材，我们利用日本北陆地方的传统捻线技术，把纤维内在的柔软性质发掘了出来，同时也使长度很长的材料能够进入施工现场，完成了这项抗震加强的工程。碳纤维的抗拉强度据说是钢铁的7倍，而且不会因温度变化而伸缩，不需要后续紧固，作为建筑材料拥有巨大的潜力。

Carbon fiber was used to seismically strengthen an existing concrete office building. A traditional twisting technique used locally in Hokuriku was applied to carbon fiber which has only be utilized as a hard rod-shaped material up until now in order to restore the inherent flexibility of carbon fiber. This enabled long fibers to be brought into the site to perform the seismic strengthening work. Carbon fiber which is said to have seven times the tensile strength of steel does not expand or contract due to heat, eliminating the necessity of retightening afterwards, and providing huge potential as a construction material.

在地下埋入钢构件，地面上只看到轻盈透明的碳纤维线。

Steel members were embedded in the ground so that only the light transparent carbon fiber strands above the ground are visible.

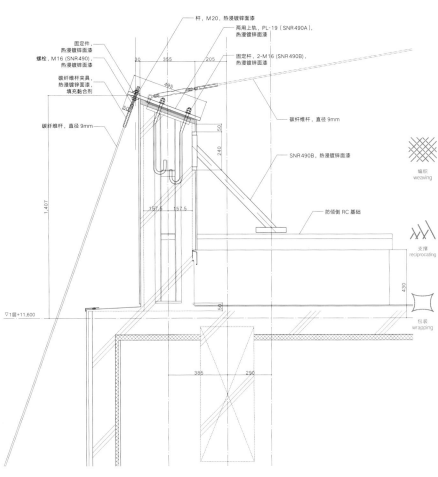

杆，M20，热浸镀锌面漆

两用上轨，PL-19（SNR490A），热浸镀锌面漆

固定杆，2-M16（SNR490B），热浸镀锌面漆

固定件，热浸镀锌面漆

螺栓，M16（SNR490），热浸镀锌面漆

碳纤维杆夹具，热浸镀锌面漆，填充黏合剂

碳纤维杆，直径 9mm

碳纤维杆，直径 9mm

SNR490B，热浸镀锌面漆

防倾倒 RC 基础

▽1层+11,600

编织
weaving

支撑
reciprocating

包裹
wrapping

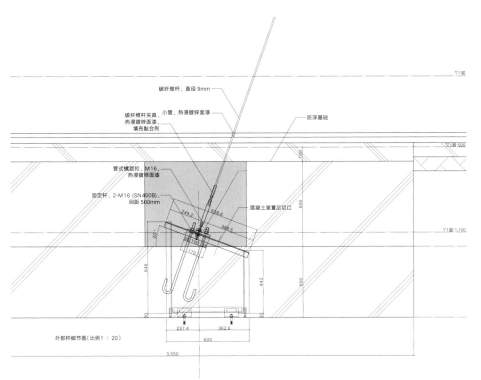

▽1层

碳纤维杆，直径 9mm

碳纤维杆夹具，小管，热浸镀锌面漆
热浸镀锌面漆，填充黏合剂

防浮基础

▽1层-500

管式螺旋扣，M16，
热浸镀锌面漆

固定杆，2-M16（SN400B），
间距 500mm

混凝土装置后切口

▽1层-1,100

▽1层-1,100

网格
grid

外部杆细节图（比例1∶20）

3,550

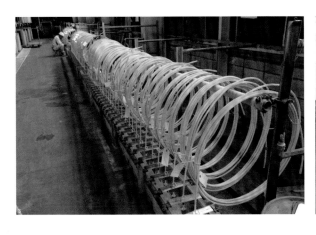

編織
weaving

支撐
reciprocating

包裝
wrapping

网格
grid

品川新站（暂名）
New Shinagawa Station

位置：日本东京

设计时间：2016年至今

估计建造时间：2020

主要用途：车站

总建筑面积：7,600m²

建筑施工：东日本铁路公司（East Japan Railway Company）、JR东日本顾问公司（JR East Consultants Company）、JR东日本设计公司（JR East Design Corporation）

建筑设计：隈研吾建筑都市设计事务所（Kengo Kuma and Associates）

钢木混合结构的框架上覆有特氟龙膜材，形成一个半透明的大屋顶，有日本纸窗般的效果。明快轻盈的大屋顶让新时代的车站建筑与城市更好地融合在一起。

大屋顶有着折纸般的几何形态，仿佛日本传统村落连绵的屋顶，体现出人文关怀。屋顶的间隙插入透明的ETFE膜，减轻特氟龙屋顶的封闭感，让人能够看到天空，感受到阴晴变化。

A large roof with an effect similar to a Shoji screen was achieved by covering a mixed structure consisting of steel and wood with a Teflon translucent membrane. The concept for this light and bright large roof was to create a station building for a new age where there is a seamless flow between the station building and surrounding city.

Origami style geometry was used to divide the large roof in order to create a look and human scale like the succession of roofs in a Japanese village. Transparent ETFE membranes were inserted into the gaps between roof sections in order to alleviate the cooped up feeling of the Teflon membrane roof as part of a challenge to create a new membrane structure in which you can see the sky and feel the weather.

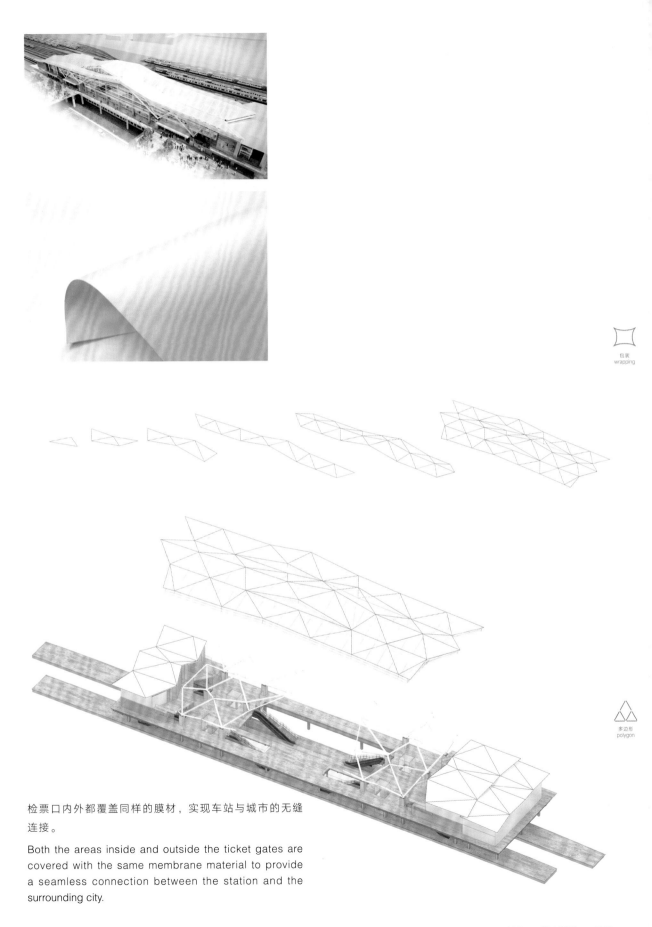

包装
wrapping

多边形
polygon

检票口内外都覆盖同样的膜材，实现车站与城市的无缝连接。

Both the areas inside and outside the ticket gates are covered with the same membrane material to provide a seamless connection between the station and the surrounding city.

包装
wrapping

多边形
polygon

简历
Profile

隈研吾

出生于1954年，东京大学建筑学科硕士。1990年成立隈研吾建筑都市设计事务所。现任东京大学教授。

1964年，丹下健三为东京奥运会设计了代代木国家体育馆。少年隈研吾为体育馆建筑所感动，立志成为一名建筑师。

在东京大学求学时，师从原广司、内田祥哉。研究生时期，穿越非洲撒哈拉沙漠进行村落调研，感受到了村落的美与力量。

在哥伦比亚大学做访问学者归来后，1990年在东京设立隈研吾建筑都市设计事务所。

此后在20多个国家做过建筑设计，曾获得"日本建筑学会奖"、芬兰"自然木造建筑精神奖"、意大利"国际石材建筑奖"等诸多奖项。

隈研吾追求融入当地环境和文化的建筑，以人性化的尺度、柔和细腻的设计见长。此外，通过探索可以取代钢筋混凝土的新型建筑材料，寻求工业化社会之后建筑发展的方向。